Prediction Technologies for Improving Engineering Product Efficiency

Lev M. Klyatis

Prediction Technologies for Improving Engineering Product Efficiency

 Springer

Lev M. Klyatis
Professor Emeritus
Habilitated Dr.-Ing., Dr. of Technical Sciences, PhD.
Jersey City, NJ, USA

ISBN 978-3-031-16657-0   ISBN 978-3-031-16655-6   (eBook)
https://doi.org/10.1007/978-3-031-16655-6

This Springer imprint is published by the registered company Springer Nature Switzerland AG
The registered company address is: Gewerbestrasse 11, 6330 Cham, Switzerland

To my wife Nellya Klyatis
L. M. K.

Preface

The book consists of two basic components:

A. Updated essence of the author's previous eight books (total over 2500 pages), including one book in Russian, one book in Chinese, and following six in English:

- *Trends in Development Accelerated Testing for Automotive And Aerospace Engineering*. Elsevier (Academic Press)
- *Reliability Prediction and Testing Textbook*. John Wiley & Sons.
- *Successful Prediction of Product Performance:*
  Quality, Reliability, Durability, Safety, Maintainability, Life-Cycle Cost, Profit, and Other Components. SAE International.
- *Accelerated Reliability and Durability Testing Technology*. John Wiley & Sons.
- *Accelerated Quality and Reliability Solutions*. Elsevier.
- *Successful Accelerated Testing*. Mir Collection.

B. New material

Our society includes many directions of people's activity. Each of them has positive and negative aspects. Whoever is involved in processing civilization's development needs to take into account all of these factors. Therefore, anyone involved in the development of any field of activity needs to do it in the following common steps.

- The first step is careful studying of real-world conditions for subject of activity. The final success of this studying depends on how deeply the problems were studied
- The second step is studying the negative aspects of trends and the causes of a problem problem's development. Here the investigator highlights the danger—illuminating the consequences of this cause. This means differentiating between problem symptoms and problem fundamental causes. For example, in developed countries, one often reads the reasons for equipment recalls, especially in automotive industry, are low safety and reliability. In fact, these are results, symptomatic consequences, and not reasons. In fact, the reason is

low level of efficiency prediction during design and manufacturing. Therefore, often the actions for increasing safety and reliability do not lead to decreasing recalls. We will see in this book how the instances of recalls are increasing through the years.

– The third step is eliminating any negative aspects and their trends in the product's development. The effectiveness of this step depends on how well the actions to eliminate negative trends for the product's development were studied.
– The fourth step is developing methodology for accurate simulation of real-world conditions.
– The fifth step is developing methodology and equipment for accelerated reliability and durability testing technology as a source for obtaining initial information for practical successful prediction of product's efficiency.
– The sixth step is developing methodology for successful prediction of product efficiency during research, design, and manufacturing.
– The seventh step is providing successful prediction of product's efficiency.
– And the eighth step is practical realization of prediction technologies during any time of the design and manufacturing for improving engineering product efficiency.

If people would follow the above steps correctly, they could avoid many mistakes during their research activities. This relates to all new technologies and basic scientific endeavors.

Following them will further the development of prediction technologies that enhance engineering product efficiency.

Additionally, the strategies described here can be useful for other areas of problems solving research. For example, prediction in engineering efficiency is a very important component of economics. A successful prediction efficiency in engineering includes results in lower life-cycle product costs and other benefits.

Often professionals whose work is related or involved with prediction, focus on its components and do not take into account that prediction is a complex problem. And for achieving successful practical solutions, it is necessary to integrate every factor that is active in the problem. Otherwise, true causes and effects cannot be determined with accuracy. This failure will always affect the efficiency (and it's components) of the product's final operations. Unfortunately, current standards and practices, especially in engineering do not take this need for systemic integration into account or minimally take it into account.

The author has been directly involved in simulation, testing, and prediction engineering over 60 years. During this time his effort has been connected with numerous testing process areas. Among them are research, design, and manufacturing of new technologies leading to the production of innovative products that lead to the betterment of human civilization. Analysis of developing trends in these areas over many years, as well as studying the work of other professionals has helped him to study prediction technologies for improvement engineering product efficiency.

From this research came numerous publications. Among them were *Trends in Development of Accelerated Testing for Automotive and Aerospace Engineering* and

Reliability Prediction and Testing Textbook, as well as other books, articles, and papers, and inventions of new methods and tools. His publications demonstrated how improvements in the design and manufacturing of products could be achieved through effective product testing. By contrast, low level of prediction technologies occur when industrial engineers or academics, who are involved in engineering product efficiency, do not have accurate and complete systematic initial information for their product's successful prediction.

This book will show how this statement is true. The goal of this book is eliminating the gap for the above problem solution. Therefore, this book includes analysis of current technologies for improving engineering product efficiency, as well as the fundamentals of new direction of prediction development – successful prediction of product efficiency, and relates to different types of engineering, including civil, electronic, aerospace, aircraft, automotive, and others. It includes:

- Analysis of damage for a product's population and economic costs from usage of unsuccessful methods of efficiency prediction during the research, testing, design, and manufacturing stages of the product
- Analysis of why current prediction technologies are not effective
- The essence of new direction of successful efficiency prediction technology for new product during design and manufacturing stages (strategy, methods, and equipment) with elimination of ineffective methods
- Partial results of implementation of this new prediction direction
- Basic methodological aspects of successful efficiency prediction
- Basis of technology accelerated reliability/durability testing (ART/ADT) for obtaining accurate initial information for successful efficiency predictions
- Basis of accurate physical simulation of real-world conditions for providing ART/ADT and successful prediction of product efficiency.

The book does not consider testing and effectiveness of given functional process of technical devices and technologies (besides any problems that are directly connected to the described processes), because they are well known by producers and users, and included in many publications.

This book is for teachers, students, and professionals who involved in the following product and processes:

- Concepts of successful prediction
- Why the current situation with the above is not as successful as they might be
- Usage of prediction in other engineering disciplines
- Testing
- Reliability
- Durability
- Maintainability
- Supportability
- Safety
- Life cycle cost
- Reasons and decreasing recalls

- Increasing profits
- Accurate simulation of real-world conditions
- Criteria of this simulation
- Methodology of this simulation
- Providing predictions, especially successful predictions
- Criteria of successful a prediction
- Reliability prediction
- Durability prediction
- Quality prediction
- Efficiency prediction
- Prediction of necessary spare parts requirements

The author wishes to express his thanks and appreciation to Lloyd H. Muller for his English language editing of this book.

Jersey City, NJ, USA   Lev M. Klyatis

Introduction

Why Does the Book Has This Title?

There are many recent scientific and technical prediction technologies and practical applications that mostly do not fulfill their essential role for improving engineering product efficiency. This leads to decreasing the actual product efficiency in comparison with what was planned. Most publications in the testing area concentrate on evaluating their functional possibilities and do not offer information for efficiency prediction, including quality, reliability, safety, durability, maintainability, life-cycle cost, profit, and other important components during a product's service life.

Let us consider the product number of recalls problem that has plagued industry for many years. They have been increasing significantly during the last dozen years. As one reads the official literature, as well as *The New York Times, Wall Street Journal*, or other sources, we will find that the basic reasons given for recalls are decreasing safety and reliability of the product.

But these are results, not reasons (causes) of the problems. The untold reason is an unsuccessful prediction of product efficiency during its design and manufacturing stages. Therefore, increasing investigations in safety during the years has not decreased the number of recalls. Let us consider the situation with automotive vehicles. As demonstrated in Table 1, from 2000 to 2019 the recalls number in the USA increased more than two times [1].

These statistics correlates directly with the number of fatalities and lost money.

For example [1]: "… According to the World Health Organization, approximately 1.3 million people die each year because of road traffic crashes. Road traffic injuries remain the leading cause of death for children and young adults aged 5–29 years. An additional 20–50 million suffer non-fatal injuries, often resulting in long-term disabilities."

Another example:

While American's drove less in 2020 due to the pandemic, NHTSA's early estimates show that an estimated 38,680 people died in motor vehicle traffic crashes—the largest projected number of fatalities since 2007 [2].

Table 1 Increasing recall number of automotive vehicles in the USA from 2000 to 2019 [1]

Year	2000	2001	2009	2010	2011	2012	2013
Recalls (millions)	24.6	13.6	17.2	22.6	15.6	18.3	26.3
Year	2014	2015	2016	2017	2018	2019	
Recalls (millions)	60.4	86.3	76.0	42.6	35.3	53.1	

Table 2 Recall rates by manufacturer (January 1985–September 2016) [4]

Recall rates by manufacturer (Jan. 1985–Sept. 2016)				
Rank	Manufacturer	# Cars recalled	# Cars sold	Recall rate per 1000 cars
1	Porsche	392,563	739,812	531
2	Mercedes-Benz	3,664,182	5,874,888	624
3	Kia	5,169,239	6,557,319	788
4	Tesla	85,184	91,046	936
5	Mazda	8,783,819	9,201,683	955
6	General Motors	129,225,450	134,895,276	958
7	Subaru	6,648,076	6,749,209	985
8	Toyota	52,614,771	51,188,734	1028
9	Nissan	28,771,128	27,709,771	1038
10	Jaguar Land Rover	1,709,017	1,601,295	1067
11	Mitsubishi	5,418,810	4,973,757	1089
	Industry Average	**527,406,265**	**472,971,556**	**1115**
12	Ford	111,043,367	97,493,753	1139
13	Volvo	3,398,341	2,940,087	1156
14	BMW	7,730,152	6,461,488	1196
15	Hyundai	13,606,160	10,743,745	1266
16	Honda	46,117,074	35,276,710	1307
17	Chrysler (FCA)	89,647,293	63,057,833	1422
18	Volkswagen Group	13,381,639	7,415,150	1805

If we compare the same period from 2019, there is an increase of 4.6%. Also, according to NHTSA's Q1 2021 Fatality Estimates, there were 8730 deaths in motor vehicle accidents in the first three months of 2021. These figures represent a 10.5% increase from the 7900 fatalities they projected for the first quarter of 2020. The NHTSA will publish the final statistics from all of 2020 through late 2021 [2].

This leads to losses of multi-billion dollars annually. The study, "The Auto Industry's Growing Recall Problem—and How to Fix It," reported:

> ... shows that automakers and suppliers paid almost $11.8 billion in claims and recorded $10.3 billion in warranty accruals for U.S. recalls in 2016. That $22.1 billion total is an estimated 26 percent increase over the previous year [3].
>
> In addition, suppliers' share of total recall costs has tripled from 5 to 7 percent from 2007 to 2013, to 15 to 20 percent since 2013, according to the study, while "the frequency that suppliers are named in recall notices has doubled [3].

This recall problem relates to companies located in the USA, as well as Europe and Asia (Table 2).

This applies not only to engineering area of endeavor. For example, The Food Safety Inspection Service (FSIS) of the US Department of Agriculture's Announcement July 8, 2021, highlighted the recall of "8,492,832 pounds of ready-to-eat (RTE) chicken products [by Tyson Foods Inc., a Dexter, Mo. Establishment] that may be adulterated with Listeria monocytogenes." While the product was distributed to schools, it resulted from a commercial sale and was not part of food provided by the USDA for the National School Lunch Program [5].

But, returning to manufactured products. As can be seen from data recorded in 2022: "Automakers issued 1.3 million recalls for the week (Feb. 3–10т)), including a Tesla recall involving 817,143 vehicles, a Hyundai recall of 357,830 Santa Fe and Tucson sports utility vehicles and Kia is recalling 126,747 Sportage SUVs, according to recall notices poson the U.S. National Highway Traffic Safety Administration website.

Hyundai and Kia are telling the owners of the affected vehicles to park them outdoors because they can catch fire even if the engines have been turned off. The recalls from the two Korean automakers are another in a long string of fire and engine failure problems that have dogged the companies for the past six years" [6].

This is just a small sample of a current industry's unsuccessful prediction of product efficiency methods. Scientists, researchers, and designers are intending one result, and obtain, in fact, a different one. Dreaming of fresh solutions they want to move civilization forward, but often go backwards. Therefore, the investigations have to come more in development of testing area, including real-world accurate simulation.

For another example, many new developments in electronics have led to fewer thinking humans using their brains (electronic think for them). This leads them to be, first, zombies, and then animals, where the basic difference greater human possibilities to think should lie. Often you can see now that people involved in front desk of medical offices or in stores cannot calculate a simple sum, for example, 10 ± 6 without calculator. On a more advanced level, a comparable situation was in the nuclear area, during the 1930s where great scientists wanted to advance civilization's beneficial powers, but they could not predict accurately that one result of their work would be a tool for mass overkill.

Moreover, as people's lives increasingly depend on electronics, the chances for failure increase because electronic devices work less reliably than people's brains. If some electronic components fail, the entire system fails. It means that the work of both electronic devices and people's work must be successfully combined. For this important combination to be achieved, successful prediction is required.

Another situation is in development artificial intelligence that has inaugurated a widening usage of autonomous vehicles. The author has analyzed many programs of international symposiums and congresses that are focused on this subject. From them, he has seen how the testing of these new systems has been mostly concentrated on the functional testing of the devices, but very little on reliability and durability testing. This failure of imagination exposes future cars to increased probability of fatal accidents under circumstances where drivers will have little or no control over their cars that they once drove manually.

Likewise, the development and implementation of electrical vehicles (EV) are progressing very quickly. In this case, again like autonomous vehicles, people are mostly studying the positive aspects of this area. But if we ignore the negative aspects of this direction, the situation will be like 120 and more years ago, when there were developed electrical cars and other vehicles such as the Stanley Steamer (Fig. 1) [7], but not accepted widely by the public. The reason for this potential failure, together with business problems, is the same: poor testing for the product's efficiency.

Many other similar situations have existed in other areas of people's lives, where successful development and long-time usefulness were very important to them and new inventions that had promised failed. All of these examples make the same point. Proper applications of resources and enlightened effort for engineering accurate prediction are needed. This book will highlight more of these failures and how proper testing methodologies, especially real-world simulation, could have averted them.

Long-term scientific efficiency prediction in engineering is a component of many areas of people's activity and has common strategic and methodological aspects. Therefore, it is very important for life and the development of humanity that effective simulation and testing be conducted on the inventions intended for human usage.

Of course, the author's work cannot envelop all varieties of problem solution, but a powerful system of defining problems, following basic directions for finding

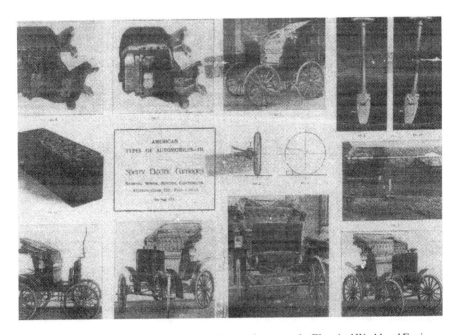

Fig. 1 Sperry electric automobile, illustrated in a supplement to the Electrical World and Engineer, July 22, 1899 [6]

solutions, and implementing them brings positive benefits for human society. Therefore, it is necessary for effective prediction techniques to be developed further.

Of course, problems occurring from poor product' efficiency lead to tragic accidents and human suffering.

Unfortunately, current trends of poor product efficiency are increasing from year to year [4, 8, 9]. The basic reason is simple. Professionals are analyzing and working mostly with results, like low safety and reliability, but not with their causes. Specifically, poor testing leads to poor prediction of product efficiency [10–13].

The consequence of this situation is why successful prediction of development product efficiency is so important for people's lives and civilization's development.

This book demonstrates how one can overcome the causes of current ineffective predictions by improving their methodology; instituting complete, integrated simulation; and testing real-life influences on products during their research, design, and manufacturing stages of development.

References

[1] Auto Accidents, Deaths, Injuries, & The Statistics Behind The Legislation. https://carsalesbase. com/auto-accidents-deaths
[2] 2020 Fatality Data Show Increased Traffic Fatalities...[https://www.nhtsa.gov/press-releases/2020-fatality...]
[3] Wagner I. (2021) Motor vehicles and equipment recalled in the United States from 2000 to 2019 (in millions). Statista. March 30, 2021
[4] ANISA LIBRELL (2018) Auto recall bill grew 26% to $22 billion in 2016, study says. Automotive News. January 30, 2018
[5] WASHINGTON, July 3, 2021—Tyson Foods Inc. Recalls Ready-To-Eat Chicken Products Due to Possible Listeria Contamination
[6] Kia, Hyundai and Tesla have recalled more than 1.3 million cars this week. See the recall list. U.S. National Highway Traffic & Safety Administration USA TODAY. CARS. Published Feb. 11, 2022 Hyundai recall: Santa Fe, Kia Sportage, Tesla cars recalled this week (usatoday.com).
[7] Thomas Parke Hughes (1971) ELMER SPERRY. Inventor and Engineer. The John Hopkins University Press. Baltimore and London
[8] Lev M. Klyatis (2016) Successful Prediction of Product Performance. Quality, reliability, durability, safety, maintainability, lifecycle cost, profit, and other components. SAE International.
[9] Lev M. Klyatis, Edward Anderson (2018) Reliability Prediction and Testing Textbook. WILEY
[10] Lev M. Klyatis (2020) Trends in Development Accelerated Testing for Automotive and Aerospace Engineering. Academic Press (ELSEVIER)
[11] Kam L. Wong (1990) What is wrong with existing reliability prediction methods? Quality and Reliability Engineering. September 1990.
[12] J.A. Jones and J.A. Hayes (2010) A Comparison of Electronic Reliability Prediction Methodologies. International Electronics Reliability Institute. Department of Electronic and Electrical Engineering Loughborough University of Technology. Leicestershire, LE11 3TU, United Kingdom
[13] https://www.sciencedirect.com/science/article/pii/S100093611730239X—! Diganta DAS Chuan LI Enrico ZIO Michael PECHT (2018) A Critique of Reliability Prediction Techniques for Avionics Applications. Chinese Journal of Aeronautics. Volume 31, Issue 1, January 2018.

Contents

About the Author

 **Lev M. Klyatis** is a senior adviser at SoHaR, Incorporated. He holds three doctoral degrees: PhD, Engineering Technology; ScD, Doctor of Technical Sciences (a high-level East European doctoral degree); and Engineering, Habilitated Dr.-Ing. (a high-level West European doctoral degree).His major scientific/ technical expertise has been in the development of new directions for the successful prediction of product efficiency during any given time, including reliability and durability testing technology, and accurate physical simulation of real-world conditions. This direction is based on new ideas and unique approaches that will lead to future development for improving society. This new direction was founded on the integration of real-world inputs, safety aspects, affecting human factors, improvement in the engineering culture, and accelerated product development. Developing a new methodology for reducing complaints and recalls, he formalized concepts on how to avoid the negative aspects of accelerated testing and prediction, and the misconceptions that are prevalent in today's, and especially could be tomorrow's (autonomous vehicles and digitalization) engineering. His approach has been verified in numerous industries, primarily in the automotive, farm machinery, aerospace, and aircraft fields. He has shared these methods as a consultant to Ford, DaimlerChrysler, Nissan, Toyota, Jatco Ltd, Thermo King, Black and Decker, NASA Research Centers, Carl Schenk (Germany), and many others.He was recognized as a full professor by the USSR's Highest Examination Board and subsequently taught at the Moscow

University of Agricultural Engineers.He was selected to be on the US–USSR Trade and Economic Council, the United Nations Economic Commission for Europe, the USA Representative for the International Electrotechnical Commission (IEC), and the International Standardization Organization (ISO). He also served as an expert at the International Standardization Organization and International Electrotechnical Commission (ISO/IEC) Joint Study Group in Safety Aspects of Risk Assessment. He was the research leader and chairman of the State Enterprise TESTMASH, Moscow, Russia, and principal engineer of a state test center.He is presently a member of the World Quality Council, the Elmer A. Sperry Board of Award, SAE International G-41 (former G-11) Reliability Committee, and the Integrated Design and Manufacturing Committee of SAE International World Congresses. Dr. Klyatis has been an SAE Fellow, Session Chairman for SAE World Congresses in Detroit since 2012. He has been a seminar instructor for the American Society for Quality.Lev M. Klyatis is the author of over 300 publications, including 12 books, and holder of more than 30 patents worldwide. He frequently speaks at technical and scientific events that are held around the world.

Chapter 1
How Was Began Development of New Direction "Successful Prediction of Engineering Product Efficiency"

The Beginning of Problem-Solving

The author, after graduating from Ukraine's Dnepropetrovsk Agricultural University (Department of Agricultural Mechanization) as a mechanical engineer, obtained a position as a test engineer at the Ukrainian State Test Center, which was located in a village, 50 miles from Kyiv, capital of Ukraine. After learning the work of this organization, he was surprised by its undeveloped testing system. In college he was more of an athlete (Fig. 1.1) than a student, and therefore, his brain was not frozen with traditional engineering theories where machinery and field conditions were considered mostly as stationary processes (not dynamic processes) that do not change during any time of machinery usage.

He did not obtain a traditional knowledge in engineering disciplines but he was guided by his passion for understanding how things actually worked that helped him to recognize the real life processes related to machinery testing. But from the beginning, his knowledge about engineering fundamentals were poor. During the first meeting, the principal engineer of the Ukrainian State Test Center said that Lev Klyatis was a terrible test engineer, but thankfully he will soon move to Kyiv, to live with his family. The author replied himself: "Well, I will leave this test center, but afterward, you will say that I am the best test engineer."

The principal engineer did not understand how sport (Fig. 1.1) strengthens a person's character and his will to victory and how multiple champion the author belonged to this category. The author continued to be involved in sport. First, he began running in the evening after work. The principal engineer who saw him said that people in the village smiled at how a test engineer would run in his underpants. Stop doing this was his advice. After that the author found a room and began training as a weight lifter. The next year he became a Kyiv state champion and was invited to train for the Ukrainian championship. During his training time, he did not

Supplementary Information The online version contains supplementary material available at https://doi.org/10.1007/978-3-031-16655-6_1.

L. M. Klyatis, *Prediction Technologies for Improving Engineering Product Efficiency*, https://doi.org/10.1007/978-3-031-16655-6_1

Fig. 1.1 Lev Klyatis, as a student of Dnepropetrovsk Agriculture University, receiving an award for sport achievements. In the middle, Professor Lecenko, the president of this university, is smiling

work as an engineer but still obtained his salary, as usual in the USSR. Again, the principal engineer saw him and asked: "If you attend these special training courses, who will do the testing of your scheduled machines? You have two choices: you will be an athlete, or you will continue your work as a test engineer." As a result, the author stopped competing but continued his involvement in sport. He wrote this story, because he believes and it is in fact that sport helps people to overcome many life problems, which was earlier seemed impossible.

During the following time, the author studied advanced literature on the development of testing technology. He saw how testing technology practices were poor. In the test center's protocol form for testing results was a question about "braking and other defects and their reasons." When test engineers filled out this form after testing, they usually wrote, "not enough strength" or "not hard enough." Of course, this did not help designers to eliminate the reason for defects. From this observation the author began to look for the real causes of reliability problems. For this goal he conducted additional field testing research. His test protocols were more complicated but sometimes helped to show designers how to eliminate the reasons for failures. By contrast, the testing results by state test centers did not help designers to eliminate the reasons of failures and increase reliability, durability, and other efficiency components.

The author was asked by experienced test engineers why the country lost billions from undeveloped test systems. The process he saw led to a clear answer. The farm machinery was tested one season in the field. If test results corresponded to given requirements, the new machinery (tractors, harvesters, etc.) was approved for manufacturing. Nobody—not test engineers, designers, or manufacturers knew what would happen with the reliability, durability, etc. of the tested machines during subsequent seasons of farming. Therefore, the users were offered lower reliability, quality, and maintainability and higher usage cost than what was planned during the designing and manufacturing stages. The common picture were a number of idle farm machineries throughout their planned service life (7–10 years). As a result, billions of rubles were squandered and food production was hindered.

When the author asked experienced colleagues why the quality of Western machinery was higher, the answer was they used higher-quality metals and other materials. Also, the quality of their manufacturing technology was higher. Therefore, their machinery's reliability and durability were higher. But the author noted how Western engineers also have problems similar to ours in usage areas, because, finally, their testing system was also poor. The author's question "Why" the answer was: "If you are so smart, solve the problem how one can achieve during the research and design stages of development successful predicting product efficiency during the product's service life.

The author continued to think and look for solutions to the problem of successful prediction of product efficiency. It was understandable that prediction methodology development alone cannot help (as many professionals think until now) to solve the problem. For an effective solution, one needs relevant initial information about the environment of its normal service life that can help calculate the prediction indices. However, recording this information over a planned service of 7–10 years would obviously slow a process. Instead, this information could be obtained through accelerated testing, which could obtain initial information for service life in a much shorter time.

The author learned from literature and research results that there were several types of accelerated testing which were used for different purposes. But, in general, the level of accelerated testing in Soviet Union, as well as in Western countries, was primitive. These testing methods did not simulate accurately the real-world conditions. Therefore, using their results did not offer accurate initial information for successful prediction of a product and its component efficiency. One can read more about this problem in Chaps. 2 and 3 of this book and in more detail in the cited books [1–4].

As a result, the author understood that solutions of successful prediction of product efficiency, as well as reliability, safety, durability, and other efficiency components, never existed. He saw the necessity of accelerated testing technology that could lead to a successful prediction of product efficiency. But this technology development needed an integrated team of professional specialists, not just one engineer.

The author wanted to involve his colleagues in solving this problem. But this problem was not supported by any funded testing program that would pay for this effort, and nobody wanted to work for free. This was particularly true if the work extended beyond their normal duties, for which they got their salaries.

After thinking and looking at how he could overcome this obstacle, he developed his first crucial step: carefully studying the field conditions that influence product failures. These conditions had to be simulated for accelerated testing. He wrote his PhD dissertation on this subject, and then he was invited to the Lviv State Test Center in the west of Ukraine (Soviet Union had 33 state test centers for farm machinery in different climatic regions) for the position of senior engineer (Figs. 1.2 and 1.3).

The author carefully studied world publications (translated in Russian) about the development accelerated testing. Its analysis showed that it was mostly accelerated stress testing as a type of accelerated life testing (ALT). These types of testing did not include an accurate simulation of field condition, which is the basis of successful testing. As a result, the stress testing results were different from field results. Unfortunately, this precept continues to this current time too. How one can overcome this obstacle?

The author created an idea as to how one can do it, but a solution of this problem needed a team, not just one engineer. As usual, when one wanted to solve a new idea, the help came, but from another direction that's what was expected. At this time, Mr. M. Butko, director of Lviv State Test Center (Fig. 1.4), was hired as director of Kalinin State Test Center, which was in dire situation, because their director had suffered three heart attacks and the principal engineer was more concerned attention about his cows and pigs than managing testing processes. Mr. Butko invited the author as a chair for one of the test laboratories in this organization. The author answered: "I will move to this center, but only as a principal engineer."

Fig. 1.2 Test engineer Ukrainian State Test Center Lev Klyatis on a motorcycle, which he used to provide field testing of farm machinery (with him his wife Nellya Klyatis, a medical college student in Kyiv, who visited the author one Sunday)

Fig. 1.3 Lev Klyatis—test engineer of Ukrainian State Test Center by a lake near a village in the state of Zhytomyr (Ukraine), where he provided field research and testing (shown is a moment of rest)

Fig. 1.4 Lev Klyatis, test engineer of Lviv State Test Center (first from right side), providing a seminar for farmers with a demonstration of the new field equipment for harvesting

The director replied: "You do not have experience as a manager of several people, and the principal engineer of the Kalinin State Test Center is managing hundreds of people." The author answered: "If I cannot obtain this position, I will go to the Lviv Agricultural University as an associate professor, where I was already invited."

Mr. Butko went to the Department of Agriculture of the Soviet Union (Moscow) where there were director and principal engineer positions of the state test centers of the Soviet Union. Finally, the bosses agreed to an experiment—move the young author from a simple engineering position to a principal engineering position. But the basic requirement was clear. He needed to dramatically improve the testing quality of this organization.

The author was happy. His dream of implementing his innovative ideas was possible. After obtaining this position, the author found five engineers who were ready to develop accelerated testing for free, after work time. That meant each team member needed to work overtime every day for an extra 4–5 h. This was a grueling work schedule. Of course, the author spent even more time at the center. The author infused his colleagues with his ideas of how they could dramatically improve their testing. They first built a test chamber that could simultaneously simulate multiple field influences (such as temperature, humidity, vibration, simulators of agricultural product, which were influencing the farm machinery failures) on the entire machine.

These engineers were highly qualified. One had experience in design, two had experience in building area. Others helped where needed. But the basic problem was simple: no money. People can work for free as volunteers, but materials for building must be purchased. The director, who was the chair of the financial area, did not want to help, because he thought that the author was pursuing this work for his own interest. Not to be stopped, the engineers learned thievery by stealing needed materials for building from nearby buildings at night.

The design of the testing equipment and the development methods for simulating real-world conditions were new. Therefore, from this original work, the team obtained patents and prepared numerous journal articles. The testing results demonstrated how high effectiveness from the developed methodology and equipment can be achieved.

The author (Fig. 1.5) continued to study current international developments in accelerated testing and efficiency prediction in different areas of industry and compared them with his work. He saw how his ideas were more advanced than those of leading Western countries, where professionals mostly concentrated on separate parts of the efficiency prediction problem and their simulation of separate environmental influences. From these reports he concluded their problems with product efficiency prediction were not being solved successfully, whereas his concepts could achieve impressive results.

The next problem was the implementation of these new solutions by their users. One can see in Chap. 6 of this book their first implementation results in the industrial company Bezeckselmash. Their effectiveness was recognized in Moscow through the principal designer and principal engineer of this company who lauded the author's testing ideas on improving the quality of their farm equipment.

Fig. 1.5 Lev Klyatis, principal engineer of Kalinin State Test Center (USSR), with his daughter Irina (now professor at Columbia University, New York)

After that recognition the Kalinin State Test Center in Moscow convened a meeting of principal testing engineers from all test centers of the USSR. It highlighted the author's presentation and demonstrated that practically the new accelerated testing technology worked. The result was very impressive.

Soon thereafter the government organized in the Kalinin State Test Center a special laboratory where the author and his team could further develop their theories and technology of accelerated testing of farm machinery. The team's arduous work had paid off. His team of enthusiastic engineers could concentrate their attention on this work for pay. The Kalinin State Test Center was visited by different organizations from around the country to see this miracle.

At that time, unfortunately, more conflicts erupted between the author and his director, Mr. Butko. First, the director thought that the author's ideas were just his fantasy and therefore did not want to provide any support money that he was authorized to spend. When he was invited to attend this work, he did not accept.

Later, even when he understood from other people how important this work was for the country, he realized how wrong he was, but then it was too late to do anything on it. He became angry and began creating additional obstacles to kill the program. The successful development of problem solutions needed support, especially financial support from headquarters, as well as more senior professionals in mathematics, physics, chemistry, and especially system of control, etc. Lacking these vital resources delayed the continuing development of a new scientific direction. Successful Prediction of Product Efficiency. Fortunately, 8 years of the author's practical experience in the field was enough to sustain his study of the effects of real field influences on the product efficiency and how it is connected with the product reliability, quality, durability, maintainability, and other efficiency components, as well as what one needs to simulate field conditions accurately. But overcoming bureaucratic obstacles was needed for this work to be developed further.

In time, the Department of Agriculture of the Soviet Union moved from state of Moscow to Moscow City. The new secretary wanted to update the organizational structure and work style of this department. For this goal he brought in fresh blood directly from agriculture, with emphasis on getting professionals who had doctorates. The author was included in this group. When the deputy of secretary invited him to Moscow as a principal specialist in the Department of Agriculture, the author said "I agree to move, but I have one requirement: you will organize in the All-USSR Institute of Agricultural Engineering (VIM) a special laboratory for continuing my research work in successful prediction of product efficiency." However, that demand implied he would work simultaneously in two jobs, and at that time holding two positions in two different organizations was not permitted.

To solve this dilemma, the author offered to work in the Department of Agriculture full time (8 h), and after this time and during weekends, he worked as a volunteer in the new laboratory. Other laboratory staff would only work there. The deputy of secretary had visited the Kalinin State Test Center earlier and understood how important this work was for agriculture. He also committed himself that to see a contact about inviting the author to work an increased salary and have additional requirement about research support.

With this support, the author obtained a position at the Department of Agriculture, USSR, and three months later a new laboratory "Successful Prediction of Agricultural Machinery Efficiency" was opened in All-USSR Institute of Agricultural Engineering (VIM), where the author was appointed as its chief. He brought in several part-time advanced mathematicians and a corresponding system of control from the aerospace area, which was mostly focused on space engineering. Other staff of the laboratory were invited, including two graduated students from the accelerated testing laboratory at the Kalinin State Test Center.

The author knew from previous experience that real-world input influences act simultaneously and interconnectedly. His laboratory experiments solved the problem simulating this relationship accurately by overcoming the barriers for laboratory simulations of real nonstationary random effects of field influences. These new generators of nonstationary random processes were ultimately patented. Finally, there were also developed advanced theoretical aspects of accelerated reliability/durability testing and methodological aspects of successful prediction of product efficiency.

Reviewing Sciences told the author that his research results met the requirements for the highest doctoral degree awarded in Eastern Europe—Doctor of Technical Sciences (USSR had two levels of doctoral degrees: low, PhD; and high, Doctor of Sciences; approximately one ScD is to 25–45 PhD). His dissertation "Scientific Basis of Agricultural Machinery Successful Prediction" was prepared and approved by the Scientific Council of Saratov State University. But highest attestation commission (VAC) of the USSR did not accept this nomination. Experienced people said that the basic reason for this refusal was bureaucratic in nature—the author was too young (34 years old). For engineering students today, his age is now the border for accepting young people for high-level doctoral degrees.

The author was nominated the second time 14 years later and his new dissertation was accepted for the doctor of science degree. The results of this dissertation were presented at various conferences in the Soviet Union and United Nations Economic Commission for Europe in Geneva over a 2 years period (Fig. 1.6), and a final presentation was published in New York (See author's book *Reliability Prediction and Testing Textbook*, Wiley 2018). The basic results of the problem solutions were published in various journals and patented.

The author's nine graduated students obtained their PhD and moved on to different industrial companies for developing and implementing accelerated reliability

Distr.
RESTRICTED

AGRI/WP.2/113
16 January 1969

ENGLISH
Original: RUSSIAN

ECONOMIC COMMISSION FOR EUROPE
COMMITTEE ON AGRICULTURAL PROBLEMS
Working Party on Mechanization of Agriculture

ACCELERATED TESTING OF AGRICULTURAL MACHINERY

Note by the Secretariat

In accordance with the decision adopted at the fourteenth session of the Working Party on Mechanization of Agriculture (see document AGRI/298, AGRI/WP.2/111) Mr. A.F. Kononenko and Mr. L.M. Klyatis, experts from the USSR, have prepared a final draft of the report on the accelerated testing of agricultural machinery. This report, the text of which is reproduced below, will be discussed at the Working Party's fifteenth session, with a view to its approval for publication in the AGRI/MECH series.

GE.69-762

Fig. 1.6 The cover of presentation Lev Klyatis and A. Kononenko (his boss) published by Economic Commission for Europe of United Nations in Geneva

testing programs. For example, Dr. V. Krasnov became the chief of the testing department in the Central Design Bureau of the manufacturing company BobruiskSelmash in Belorussia, where he implemented accelerated reliability testing for fertilizer applicators. Dr. E. Schechtman obtained a position as senior scientist in the Institute VNIIZIVMASH in Kyiv (Ukraine), where he devised a new technology for vibration testing (see Chap. 6).

The author's advantage was connecting theoretical research with practice, which helped him to get more effective research results than those of other colleagues. Later, this advantage was one of the basic factors in promoting his career after moving to the United States (Fig. 6.1). Sadly, this was not a smooth path that he followed.

His work was stopped for 15 years because of political reasons. The author's boss Dr. Kononenko prepared doctoral dissertation for his ScD. Unfortunately, the deputy minister of the Department of Agriculture for the USSR was afraid that Dr. Kononenko would usurp his position. To block this possibility, he organized a letter from some people from the Committee of Central Communist Party Control (KPK) suggesting that some people had prepared parts of Dr. Kononenko dissertation. If this is true, this would violate the academic regulations of the Communist Party of the USSR. The above committee was very important in the Communist Party hierarchy. In this letter the author's name was highlighted as the one who had helped to prepare this dissertation. He was summoned before this committee and asked to write a short letter that he did this. The person who invited him said that if he will do it, he will obtain the highest scientific degree in USSR—academician. The author said that he did not do this and so he cannot sign this letter. As a result, Dr. Kononenko was eliminated from Communist Party membership and he lost his ministerial position.

The Communist Party Committee of the Department of Agriculture received Dr. Kononenko's dismissal notice with the signature of the committee chair Mr. Palshe who was a member of the Soviet Union Communist Party Politbureau, the highest level of authority in the country. The cover letter stated that Department of Agriculture USSR communist committee has to consider the author's responsibility for the above controversy. The committee's investigator told the party chair of the Department of Agriculture over the phone that they needed to remove the author from the Communist Party. This action meant the loss of his positions in Department of Agriculture USSR and his laboratory will be closed. The Minister of Agriculture USSR did not want to do this, and therefore the headquarters decided that the author could remain in the party if he confessed his errors. When they told him about this, he said: "I have confessed for the many millions of rubles that I brought to my country for free." But when they showed him the reports from the Communist Party Central Committee, he understood how false they were. Finally, the mistake was in his administration of the laboratory: there are three people who worked in another company for part-time work. (The author's position as a volunteer in the laboratory meant he was not responsible for the human resource actions.) Nonetheless, holding two jobs at once was prohibited and it became the basis for his discharge.

Since, the laboratory was closed, he needed to find another job (not in Department of Agriculture USSR). He had previously received many proposals for vice president positions at various companies, but now he could not find any engineering job in Moscow, because the above scandal made people afraid to accept him. His wife, an excellent doctor-pediatrician, found a job for him as a simple engineer in a chemical research center.

So, his innovative scientific direction for the successful prediction of product efficiency was frozen for many years.

The ice of his career thawed 15 years later with the advent of *perestroika* or "openness" that was the political theme of the new Soviet leader, Michael Gorbachev. At that time there were two farm machinery departments in the Soviet Union. One of them was the Department of Farm Machinery for Stock Raising and Feeding. The farm equipment being produced by this department was notoriously unreliable for users and producers. As a result, state food production goals were not being fulfilled as planned. This created both economic costs and social problems as the Soviet citizens needed food with reasonable prices.

Mr. Gorbachev invited Mr. L. Hitrun, the previous First Deputy USSR Minister of Agriculture (he was involved earlier in farm machinery), as the minister of the Department of Farm Machinery for Stock Raising and Feeding. Mr. Hitrun invited the author and said that Gorbachev told him about the urgent need for improving the quality, reliability, and durability of farm machinery. Therefore, Mr. Hitrun proposed the author to be the chief of new reliability and testing department in All-USSR Central Institute VNIICOMZG.

This welcome demand by President Gorbachev created the possibility for implementing of the author's ideas and research results. Before the author had told his superiors how his ideas and research results were leading the world. But they did not believe him. Now he obtained the possibility for demonstrating the truth of his proposal. Now he could demonstrate the truth of his ideas.

This welcome demand made possible the successful prediction of agricultural machinery efficiency. Later it was understood that the basic developed strategy and concepts related not only to farm machinery but also other areas as well. It created a new scientific direction of successful efficiency prediction, including accelerated reliability and durability testing, and, as a result, increasing quality, reliability, durability, and other efficiency components. Another developed component was a methodology of successful prediction of product efficiency. One can read the details about this problem solution continuation and implementation in Chap. 6, as well as in other author's publications.

A significant role in the development and implementation of the author's ideas was his trip to Germany with the Russian governmental delegation. First, the delegates visited the Carl Schenck company in Darmstadt which is close to Frankfurt am Main. The Schenck employees were highly impressed by author's advances in development. They saw it was better than any competitor's testing program. For example, he demonstrated and described his vibration equipment in six degrees of freedom with the system of controlling nonstationary random vibration.

In a final protocol (see Fig. 6.28 in [4]), Carl Schenck company proposed joint venture project with the author together with company, where he worked. The Russian delegation obtained many additional benefits from this partnership.

Then the delegation moved to the Heraeus Fetch company that designed and produced advanced test chambers. The author made a presentation about the results of his department's work in the development of test chambers. In a final protocol, this company also agreed to joint venture with his organization.

The chairs of Russian delegation were surprised. Earlier they heard from the author that his department's scientific-technical solutions were the most advanced in the world. Now they were hearing from the world's famous companies' professionals the same information. Moreover, these companies were very interested in working with him. On their way back to Russia, the chairs asked the author what he needed for acceleration development of his concepts. He said: "You need to organize a specific company that is devoted to accelerating the development of testing technology and reliability in our industry, as component of successful prediction of product efficiency."

When this delegation returned home, the USSR government organized a new enterprise TESTMASH, where the author was appointed as its chairman. He built the staff of this new company and started an aggressive development and implementation of his ideas.

President's Gorbachev's "perestroika" opened the path for more presentations in Western countries. The author was nominated to the US-USSR Trade and Economic Council (ASTEC), which was organized by USSR and American governments. The meetings of this council were held twice a year—one in the United States and one in the Soviet Union. In a New York Council meeting program, there were two keynote presentations from both sides—US and USSR alike. The USSR presented the author's oral paper about the basic results of his group's solutions. It was a successful conference, because afterward the author obtained many collaboration proposals.

After the author's presentation during the ASTEC meeting, Mr. Bekh, general director of KAMAZ, Inc., the largest Russian producer of trucks (over 100,000 employees), met with the author. Mr. Bekh offered him a contract for implementing a new direction of the prediction of product efficiency.

It was started and when completed, his team developed new equipment for the accurate simulation of field conditions and accelerated reliability/durability testing for trucks. These advances led to more accurate predictions of truck efficiency during their service life. One can see more details of this work in Chap. 6. The success stories related above were just the start of a new international path for improved machinery.

The author also obtained proposals from large American and European companies, as well as other Russian companies.

Many research centers also joined in these efforts. One can see one more example in Fig. 1.7.

The above information is important, because it shows how the development of an innovative product prediction technology was created, for present projects as well as those in the future. The scheme of creating idea components is shown in Fig. 1.8. Now these ideas must be further developed and implemented to advance the benefits of civilization.

Fig. 1.7 Dr. Lev Klyatis, chairman of State Enterprise TESTMASH, in the test center VNIIZIVMASH (Kiev, Ukraine) where he implemented his ideas for accelerated reliability and durability testing development

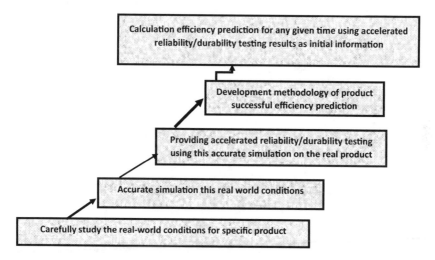

Fig. 1.8 Created principal scheme of prediction of product efficiency for any given time

Chapters 2, 3, 4, 5, and 6 of this book show the directions of these efforts must take. More detailed descriptions are found in references [1–6] along with another book in Russian and the book in Chinese.

The author's book *Trends in Development Accelerated Testing for Automotive and Aerospace Engineering* was nominated in the "52 Best Automotive Engineering Books of All Time (#12)" and "100 Best Aerospace Engineering Books of All Time (#43)" (As featured on CNN, Forbes and Inc.—Book Authority identifies and rates the best books in the world, based on recommendations by thought leaders and experts).

References

1. Lev M. Klyatis (2012) Accelerated Reliability and Durability Testing. WILEY
2. Lev M. Klyatis, Edward L. Anderson (2018) Reliability Prediction and Testing Textbook. WILEY
3. Lev M. Klyatis, Eugene L. Klyatis (2006) Accelerated Quality and Reliability Solutions. ELSEVIER
4. Lev M. Klyatis (2020) Trends in Development of Accelerated Testing for Automotive and Aerospace Engineering. Academic Press (ELSEVIER).
5. Lev Klyatis (2016). Successful Prediction of Product Performance (quality, reliability, durability, safety, maintainability, life-cycle cost, profit, and other components). SAE International
6. Lev M. Klyatis, Eugene L. Klyatis (2002). Successful Accelerated Testing. Mir Collection

Chapter 2
Analysis of Current Situation with Prediction of New Product Reliability and Efficiency

2.1 Current Methodological Aspects of New Product Reliability and Efficiency Prediction

This chapter reviews the current used methodological aspects of reliability/efficiency prediction for industries such as electronics, automotive, aircraft, aerospace, off-highway, farm machinery, and others. It then discusses why these are not successful. A detailed analysis of various aspects of reliability prediction methodologies was published in the author's book *Reliability Prediction and Testing* [1]; however, a summary of these methods will be provided in this chapter. There are many publications in reliability prediction area, because this problem is very important for the theory and practice of reliability, especially when using product reliability prediction for the provision of high efficiency of the products. Most of these publications relate to electronics and use traditional methods of reliability prediction. The chapter will end with a discussion of other methodologies that have been proven to be even more effective.

Prediction has historically been used to denote the process of applying mathematical models and data to estimate the field performance of a system (product/technology). This was done before empirical data were available for the systems being developed. Moreover, it is important for successful efficiency prediction, as will be shown below.

Traditional methods of reliability prediction utilize failure data collected from the field to estimate the parameters of the failure time distribution or degradation process. Using these data one can then estimate the reliability measures of interest, such as the expected time to failure or quantity/quality degradation during a specified time interval and the mean time to failure (or

Supplementary Information: The online version contains supplementary material available at https://doi.org/10.1007/978-3-031-16655-6_2.

L. M. Klyatis, *Prediction Technologies for Improving Engineering Product Efficiency*, https://doi.org/10.1007/978-3-031-16655-6_2

Reliability prediction models that utilize accelerated degradation data are preferred over those models using degradation data obtained under normal conditions, because often the degradation process is very slow and requires prolonged period compared to that observed under normal field conditions.

There are many uses of reliability prediction information, including components of product efficiency (life cycle cost, profit, safety, human factors influence, and others). For example, it can be used as inputs to life cycle cost or profit studies. Life cycle cost studies determine the cost of a product over its entire life. Required data include how often a unit will have to be replaced. Inputs to this process include the steady-state failure rate and the first-year multiplier.

1. Cost-effectiveness comparisons: The total cost can be divided by the cycles of operations (hours, rates of fire, etc.). The system with the lower cost would be selected for further development.
2. Over 60% of all life cycle operations are generated by operations and maintenance costs using reliability predictions to improve initial design where costs usually comprise only about 10% of life cycle costs and will reap large rewards later during the machine's operating life.
3. Scope of procedures—hardware:

   • Devices
   • Units
   • Simple serial systems
   • Design considerations for complex systems

4. Scope of procedures—software:

   • Systems
   • Modules

The Bellcore hardware reliability prediction is centered only on electronic equipment. It provides predictions at the device level, at the unit level, and for simple serial systems, but it is primarily aimed at line replaceable units (LRU). These units are replaceable non-repairable assemblies or plug-ins. The goal is to provide the user with useful information on how often units will fail and have to be replaced.

The software prediction procedure estimates software failure intensity. It applies to systems and modules.

The Bellcore reliability prediction procedure consists of three methods:

1. Parts count—The predictions are based solely on adding together the failure rates for all the devices. This is the most commonly used method, because laboratory and field information that is needed for the other methods is usually not available.
2. Incorporating laboratory information—Device or unit level predictions are obtained by combining data from a laboratory test with the data from the parts count method. This allows suppliers to use their data to produce realistic predictions of failure rates and is particularly suited for new devices for which few field data are available.

3. Incorporating field information—This method allows suppliers to combine field performance data with data from the parts count method to obtain accurate reliability predictions.

Mechanical reliability prediction [13] uses various stress factors under operating conditions as a key to all proposed models for different devices. The number of such factors that must be included in any calculation may appear to be excessive, but tailoring methods can be used to remove factors that have little or no influence, or for which no data are available. This situation may occur more often in testing mechanical devices (bearings, compressors, pumps, etc.) than for electronic systems.

The problems found with all attempts to predict the reliability of mechanical systems are a lack of specific or generic failure rate data, lack of information on the predominant failure modes, and lack of information on factors which influence the reliability of the mechanical components.

The role of testing as a source of accurate initial information is a critical factor.

Misunderstanding the role of testing and definitions can easily lead to poor product and human reliability and decreased financial profits.

Prediction is useful if it helps to slow efficiency degradation and prevent the failures of the product reliability characteristics and, finally, improve the efficiency of components during usage time.

There are many recent publications that address electronics, automotive, and other product recalls. They mostly address safety matters that are related to a product's quality, reliability, and durability. Failures here cause economic problems and affect peoples' lives.

But they are not considered efficiency prediction. At the core, reliability and safety issues and other efficiency problems are the results of failures, not the causes. The actual causes of recalls and many other technical and economic problems are the inefficient or inadequate prediction of product reliability, safety, and other efficiency components during the design and manufacturing processes. The poorly executed prediction then negatively affects financial performance.

For example, Atiyeh and Blackwell state: "So far, about 12.5 million suspect Takata inflators have been fixed of the roughly 65 million inflators (in 42 million vehicles) that will ultimately be affected by this recall, which spans 19 automakers. Carmakers and federal officials organizing the response to this huge recall insist that the supply chain is churning out replacement parts, most of which are coming from companies other than Takata. For those who are waiting, HTSA advises that people not disable the airbags; the exceptions are the 2001–2003 Honda and Acura models that we listed on this page on June 30, 2016—vehicles which NHTSA is telling people to drive only to a dealer to get fixed. Meanwhile, a settlement stemming from a federal probe into criminal wrongdoing by Takata is expected early next year—perhaps as soon as January—and could approach $1 billion" [3].

A key consideration to solving these problems is the use of advanced testing methods and equipment. Such procedures could be accelerated reliability testing and accelerated durability Testing (ART/ADT). These systematic procedures assure successful prediction of industrial product efficiency, including its components. Such advances in technology usually do lead to more complicated products and increasing economic development. They also require more effort to accurately simulate the real-world conditions and predict product efficiency.

However, when performed successfully, prediction is useful in the product start-up, design, and manufacturing stages of development. It touches our lives economically and provides mechanisms to assess components of professional activity for product improvement at any time interval, from the earliest stages of R&D throughout the entire product life cycle, including usage.

Currently, there are many other publications mostly related to the theoretical aspects of reliability prediction. Many primarily relate to failure analysis.

Some popular failure analysis methods and tools include the following:

- Failure reporting, analysis, and corrective action system (FRACAS)
- Failure mode, effects, and criticality analysis (FMECA)
- Failure modes and effects analysis (FMEA)
- Fault tree analysis (FTA)

FavoWeb is Advanced Logistics Development's (ALD) third-generation, web-based, and user-configurable FRACAS that captures information about the equipment or processes throughout its life cycle, from design to production, testing, and customer support.

FavoWeb has been adopted by organizations that have, for the first time, implemented a FRACAS application that seamlessly communicates with any given enterprise resource planning (ERP) system (SAP, Oracle, MFGPro, etc.), while proving a user-friendly and flexible, yet robust, failure management, analysis, and corrective action platform.

The FavoWeb FRACAS features include as follows:

- Full web-based application.
- User permission mechanism complies with International Traffic in Arms Regulations requirements.
- *Inter-method reliability* assesses the degree to which test scores are consistent when there is a variation in the methods or instruments used. This allows inter-rater reliability to be ruled out. When dealing with forms, it may be termed *parallel forms reliability* [6].
- *Internal consistency reliability* assesses the consistency of results across items within a test [6].

A test that is not perfectly reliable cannot be perfectly valid, either as a means of measuring attributes of a person or as a means of predicting scores on a criterion. While a reliable test may provide useful valid information, a test that is not reliable cannot be valid [7].

Now there are many publications and webinars that relate to simulation for safety purposes. They have usually demonstrated narrow thinking when attempting to solve the simulation and testing problems, because they consider only simulation of some from many real-world factors. For example, the author received from SAE International the following invitation:

"Dear Lev,

The world is riding a wave of commercial transportation innovation with the advent of the eVTOL industry. This is causing a flood of R&D funding into new and existing aerospace engineering companies. No matter the size of these companies, their new aircraft must be certified for safety by the FAA and EASA, which is a rigorous process that requires a vast amount of engineering simulation and testing.

"Technical Webinar
Certification of eTOL Aircraft:
Leveraging Simulation for Safety
and Structural Integrity"

In this 60-min webinar, two experts discuss the certification of eVTOL aircraft and how to leverage engineering simulation to ensure safety and structural integrity. With every new aircraft comes sophisticated technology that must be tested and certified as safe for the public. Examples from the past include the first delta wing fighter jet, the first supersonic airplane, and the first four-engine, two-story jumbo jet. Today's new eVTOL aircraft are no different. Most eVTOL aircraft companies are considering new combinations of tilt-rotors, tilt-wings, all-composite fuselages, fly-by-wire control systems, and rechargeable electric battery systems that have never been seen in the world of aviation. Engineers need huge amounts of simulation and testing to raise these new technologies to the levels needed for eVTOL aircraft to gain widespread adoption.

This 60-min webinar discusses the following:

- An overview of eVTOL regulatory requirements and the need to certification
- How to leverage simulation to ensure aircraft safety and structural integrity
- Simulation success stories from leading aerospace companies

Live Presentation:

Wednesday, April 6, 2022, at 2:00 pm US EDT
Speakers:

Bobby Cook, Ph.D., Industry Director, Aerospace & Defense Hexagon

James Pura, Product Marketing Manager, Hexagon

2.1.1   General Model

In practice, testing measures are never perfectly consistent. Theories of test reliability have been developed to estimate the effects of inconsistency of the accuracy of measurement. The basic starting point for almost all theories of test reliability is the idea that test scores reflect the influence of two sorts of factors [7]:

1. *Factors that contribute to consistency*—stable characteristics of the individual or the attribute that one is trying to measure.
2. *Factors that contribute to inconsistency*—features of the individual or the situation that can affect test scores but have nothing to do with the attribute being measured.

   These factors include as follows [7]:
   - Temporary, but general, characteristics of the individual—health, fatigue, motivation, emotional strain.
   - Temporary and specific characteristics of the individual—comprehension of the specific test task, specific tricks or techniques of dealing with the particular test materials, fluctuations of memory, attention, or accuracy.
   - Aspects of the testing situation—freedom from distractions, clarity of instructions, interaction of personality, sex, or race of examiner.
   - Chance factors—luck in selection of answers by sheer guessing, momentary distractions.

The goal of estimating reliability is to determine how much of the variability in test scores is due to errors in measurement and how much is due to variability of true scores [7].

2.1.2   Classical Test Theory

The goal of classical test theory is to estimate errors in measurement and to suggest ways of improving tests so that these errors are minimized. The central assumption of reliability theory is that measurement errors are essentially random. This does not mean that errors arise entirely from random processes. For any individual, an error in measurement is not a completely random event. However, across a large number of individuals, the causes of measurement error are assumed to be so varied that measure errors act as random variables.

If errors have the essential characteristics of random variables, then it is reasonable to assume that errors are equally likely to be positive or negative and that they are not correlated with true scores or with errors on other tests.

Unfortunately, there is no way to directly observe or calculate the *true score*, so a variety of methods are used to estimate the reliability of a test.

Some examples of the methods used to estimate reliability include test-retest reliability, internal consistency reliability, and *parallel test reliability*. Each method approaches the problem of accounting for the source of error in the test somewhat differently.

2.1.3   Estimation

The goal of estimating reliability is to determine how much of the variability in test scores is due to errors in measurement and how much is due to variability in true scores. There are several methods which have been developed that provide workable methods of estimating test reliability, including the following [7]:

Test-Retest Reliability Method   Directly assesses the degree to which test scores are consistent from one test administration to the next. It involves:

- Administering a test to a group of individuals.
- Readministering the same test to the same group at some later time.
- Correlating the first set of scores with the second set of scores.

The correlation between scores on the first test and the scores on the retest is used to estimate the reliability of the test using the Pearson product-moment correlation coefficient.

See also *item-total correlation* in Ref. [7].

Parallel Forms Method   The key to this method is the development of alternate test forms that are equivalent in terms of content, response processes, and statistical characteristics. For example, alternate forms exist for several tests of general intelligence, and these tests are generally seen as equivalent [7].

There are many situations in which one needs to make a prediction about a product's efficiency before the product is in production. This means prediction is needed prior to production or warranty data being available for analysis. Many companies have product development programs that require design engineers to produce designs that will meet a certain reliability goal before the project is permitted to move on to the following phases (building prototypes, pre-manufacturing, and full manufacturing). This is to avoid committing the business to investing significant resources to a product with unproven efficiency before leaving the design stage. This is especially difficult because a new design can involve components or subsystems that have no earlier testing and have no history of being used in the field by customers. Often, they encompass totally new items and not redesigned components or subsystems of existing components which would have prior histories [7].

In other cases, companies may not have the capabilities, resources, or time to test certain (non-crucial) components/subsystems of a system but still need to use some estimates of the failure rate or other characteristics of efficiency components to complete their system efficiency analysis.

Lastly, manufacturers must often submit reliability predictions usually based on a specific prediction standard with their bid or proposal for a project.

The advantages of using standards-based prediction are the following:

- They can help to complete the system reliability block diagrams (RBDs) or FTAs when data for certain components/subsystems within the system are not available.
- It is sometimes accepted and/or required by government and/or industry contracts for bidding purposes.

The disadvantages of using standards-based prediction are the following:

- Reliance on standards that may not reflect the product's actual efficiency.
- Although standards-based reliability prediction addresses prediction under different usage levels and environmental conditions, these conditions may not accurately reflect the product's actual application.
- Some of the standards are old and/or have not been updated level to reflect the latest advances in technologies.
- The result from such predictions is a constant failure rate estimation that can often be used within the context of an exponential reliability model. This is not necessarily accurate for all components and certainly not for most mechanical components. In addition, certain aspects of reliability analysis, such as preventive maintenance analysis and burn-in analysis, cannot be performed on components/subsystems that follow the exponential distribution.

So, the basic negative aspect of this approach to reliability prediction is that it is not reflective of the product's actual efficiency. Therefore, reliability/efficiency prediction results may be very different from field results, and, as a final result, the reliability/efficiency prediction will be unsuccessful.

2.1.4   Reliability Prediction for Mean Time Between Failures

Reliability prediction tools such as the ITEM ToolKit are essential when the reliability of electronic, mechanical, electrical, and other components, systems, and projects is critical for life safety. Certain products and systems developed for commercial, military, or other applications often need absolute ensured reliability and consistent efficiency. However, electronics, mechanical, and other products, systems, and components are naturally prone to eventual breakdown owing to any number of environmental variables, such as heat, stress, moisture, and moving parts. The main question is not if there will be failures, but "when?"

2.1.5 Reliability Software Overview

The reliability software modules of the ITEM ToolKit [7] supply a user-friendly interface that allows one to construct, analyze, and display system models using the module's interactive facilities. Building hierarchies and adding new components could not be easier. Toolkit calculates the failure rates, including mean time between failures (MTBF), associated with new components as they are added to the system, along with the overall system failure rate. Project data may be seen simultaneously via a grid view and a dialog view, allowing predictions to be performed with a minimum of effort.

Each efficiency prediction module is designed to analyze and calculate component, subsystem, and system failure rates in accordance with the proper standard. After the analysis is complete, ITEM ToolKit's integrated environment comes into its own with powerful conversion facilities to transfer data to other reliability software modules. For example, you can transfer your Military Handbook 217 (MIL-217) project data to FMECA or your Bellcore project to RBD. These powerful features transfer as much of the available information as possible, thereby saving valuable time and effort [7].

The following is an interesting statement from ReliaSoft's analysis of the current situation in reliability prediction [8]. To obtain high product reliability, consideration of reliability issues should be integrated from the very beginning of the design phase. The objective of reliability prediction is not limited to predicting whether reliability goals, such as MTBF, can be reached. It can also be used for as follows:

- Identifying potential design weaknesses.
- Evaluating the feasibility of a design.
- Aiding in business decisions such as budget allocation and scheduling.
- Prediction of the reliability/efficiency of electronic products requires knowledge of the components, design, manufacturing process, and, especially, expected operating conditions.
- Several different approaches have been developed to achieve reliability prediction of electronic systems and components. Each approach has its advantages but mostly disadvantages.

Among these approaches, three main categories are often used within government and industry:

- Empirical (standard bases)
- Physics of failure
- Life testing

An overview of all three approaches can be found in ReliaSoft's article "Reliability Predictions Method for Electronic Products" [8].

2.1.5.1 MIL-HDBK-217 Predictive Method

MIL-HDBK-217 is well known in military and commercial industries. Version MIL-HDBK-217F was released in 1991 with two revisions from the earlier editions.

The MIL-HDBK-217 predictive method consists of two parts: one is known as the *parts count* method and the other is called the *part stress* method [8]. The parts count method assumes typical operating conditions of part complexity, ambient temperature, various electrical stresses, operation mode, and environment (called *reference conditions*).

The failure rate for a part under the reference conditions is calculated as

$$\lambda_{b,i} = \sum_{i=1}^{n} i = \left(\lambda \ \text{ref} \right) i$$

where λ is the failure rate under the reference conditions and i is the number of parts.

Since the parts may not operate under the reference conditions, the real operating conditions may result in failure rates different than those given by the "parts count" method. Therefore, the part stress method requires knowledge of the specific part's complexity, application stresses, environmental factors, and so on. These adjustments are called *Pi factors*. For example, MIL-HDBK-217 provides many environmental conditions, expressed as πE, ranging from "ground benign" to "cannon launch." The standard also supplies multilevel quality specifications.

2.1.5.2 Bellcore/Telcordia Predictive Method

Bellcore was a telecommunication research and development (R&D) company that provided joint R&D and standards setting for AT&T and its co-owners. Bellcore was not satisfied with the MIL-HDBK-217 methods for application with their commercial products, so Bellcore designed its own reliability prediction standard for commercial telecommunication products.

Later, the company was acquired by Science Applications International Corporation (SAIC) and the company's name was changed to Telcordia. Telcordia continues to revise and update the Bellcore standard. Presently, there are two updates: SR-332 Issue 2 (September 2006) and SR-332 Issue 3 (January 2011), both titled "Reliability Prediction Procedure for Electronic Equipment" [5].

The Bellcore/Telcordia standard assumes a serial model for electronic parts and it addresses failure rates at both the infant mortality stage and at the steady-state stages using Methods I, II, and III. Method I is similar to the MIL-HDBK-217F parts count and part stress methods, providing the generic failure rates and three part stress factors: device quality factor πq, electrical stress factor πS, and temperature stress factor πT. Method II is based on combining Method I predictions with data from laboratory tests performed in accordance with specific SR-332 criteria. Method

III is a statistical prediction of failure rate based on field tracking data collected in accordance with specific SR-332 criteria. In Method III, the predicted failure rate is a weighted average of the generic steady-state failure rate and the field failure rate.

The basic disadvantage of the above approach is using only laboratory test, which does not reflect all real-world conditions. This leads to reliability and efficiency inaccurate prediction.

Chapters 3 and 4 discuss how one can improve this situation.

2.1.5.3   Discussion of Empirical Methods

Empirical prediction standards have been used for many years, and it is wise to use them with caution. The advantages and disadvantages of empirical methods have been discussed in a brief summary from publications in industry, military, and academia [43].

Advantages of empirical methods:

1. Easy to use, and a lot of component models exist.
2. Relatively good performance as indicators of inherent reliability.

Disadvantages of empirical methods:

1. A large part of the data used by the traditional models is out-of-date.
2. Failure of the components is not always due to component-intrinsic mechanisms but can be caused by the system design.
3. The reliability prediction models are based on industry-average values of failure rate, which are neither vendor specific nor device specific.
4. It is hard to collect real-world (field and manufacturing) data, which are needed to define the adjustment factors, such as the Pi factors in MIL-HDBK-217 [43].

2.1.6   Physics of Failure Methods

In contrast to empirical prediction methods that are based on the statistical analysis of historical failure data, a physics of failure approach is based on the understanding of the failure mechanism and applying the physics of failure model to the data. Several popularly used models are discussed next.

2.1.6.1   Arrhenius's Law

One of the earliest acceleration models predicts how the time to failure of a system varies with temperature. This empirically based model is known as the *Arrhenius equation*.

Generally, chemical reactions can be accelerated by increasing the system temperature.

Since it is a chemical process, the aging of a capacitor (such as an electrolytic capacitor) is accelerated by increasing the operating temperature.

2.1.6.2   Black Model for Electromigration

Electromigration is a failure mechanism that results from the transfer of momentum from the electrons, which move in the applied electric field, to the ions, which make up a lattice of the interconnect material. The most common failure mode is "conductor open." With the decreased structure of integrated circuits (ICs), the increased current density makes this failure mechanism important in IC reliability.

At the end of the 1960s, JR Black developed an empirical model to estimate the mean time to failure (MTTF) of a wire, taking electromigration into consideration, which is now generally known as the *Black model*.

2.1.6.3   Discussion of Physics of Failure Methods

A given electronic component will have multiple failure modes, and the component's failure rate is equal to the sum of the failure rates of all modes (i.e., humidity, voltage, temperature, thermal cycling, etc.). The authors of this method propose that the system's failure rate is equal to the sum of the failure rates of the components involved. In using the aforementioned models, the model parameters can be determined from the design specifications or operating conditions. If the parameters cannot be decided without conducting a test, the failure data obtained from the test can be used to get the model parameters. Software products such as ReliaSoft's ALTA program can help analyze the failure data. For example, to analyze the Arrhenius model, the life of an electronic component is assumed to be affected by temperature. The component is tested under the temperatures of 406, 416, and 426 degrees Kelvin ($^\circ$ K). The usage temperature level is 400 K.

The Arrhenius model and the Weibull distribution are used to analyze the failure data in ALTA.

Advantages of physics of failure methods:

1. Modeling of potential failure mechanisms based on the physics of failure.
2. During the design process, the variability of each design parameter can be determined.

Disadvantages of physics of failure methods:

1. The testing conditions do not accurately simulate the field conditions.
2. There is a need for detailed component manufacturing information, such as technology, material, process, and design data.
3. Analysis is complex and could be costly to apply.

4. It is difficult (almost impossible) to assess the entire system.

These limitations generally do not lead to a practical useful methodology.

2.1.7   Life Testing Method

Time to failure data from life testing may be incorporated into some of the empirical prediction standards (i.e., Bellcore/Telcordia Method II) and may also be necessary to estimate the parameters for some of the physics of failure models. But the term *life testing method* should refer specifically to a third type of approach for predicting the reliability/efficiency of electronic products.

With this method, a test is conducted on a sufficiently large sample of units operating under normal usage conditions. Times to failure are recorded and then analyzed with an appropriate statistical distribution in order to estimate reliability and other efficiency component metrics such as the B10 life. This type of analysis is often referred to as *life data analysis* or *Weibull analysis.*

ReliaSoft's Weibull++ software is a tool for conducting life data analysis. As an example, suppose that an IC board is tested in the lab and the failure data are recorded. But failure data during long period of use cannot obtained, because accelerated life testing (ALT) methods are not based on accurate simulation of the field conditions.

Conclusions
In the ReliaSoft article [8], three approaches for electronic reliability prediction were discussed. All of them are related to efficiency prediction results. The empirical (or standards based) method, which is close to the theoretical approach, can be used in the predesign stage to quickly obtain a rough estimation of product efficiency/reliability. The physics of failure and testing methods can be used in both design and production stages. When using the physics of failure approach, the model parameters can be determined from design specifications or from test data.

But when employing the life testing method, since the failure data, the *prediction results usually are not more accurate than those from a general standard model.*

For these reasons, the traditional approaches to prediction are often unsuccessful when used in industrial applications and especially cannot be used for successful efficiency prediction.

Another important reason is these approaches are not closely connected with the system of obtaining accurate initial information for calculating reliability/efficiency prediction during any period of use.

The limitations of models, methods, procedures, algorithms, and programs are outlined. The treatment of maintained systems is designed to aid workers in analyzing systems with more realistic assumptions. FTA, including the most recent developments, is also extensively discussed.

Failures have been a problem since the very first computer. Components burned out, circuits shorted or opened, solder joints failed, pins were bent, and metals reacted unfavorably when joined. These and countless other failure mechanisms have plagued the computer industry from the very first circuits to today.

As a result, computer manufacturers realize that reliability predictions are very important to the management of their product's profitability and life cycle. But there is a problem with successfulness of this prediction, because it is not based on accurate initial information for this prediction. Therefore, one has to be careful with using the above approaches for successful efficiency prediction. For more detailed description, one can read [1].

While these traditional approaches can offer a theoretically effective means of testing, they are unable to reflect or account for actual product efficiency. Its components during life cycle of a product change in accordance with usage interaction and outside effects of real-world inputs. For these reasons, these approaches are not often successful.

Failure of the Old Methods
Today, we find that the old methods of predicting reliability in electronics have begun to fail us. MIL-HDBK-217 has been the cornerstone of reliability prediction for decades. But MIL-HDBK-217 is rapidly becoming irrelevant and unreliable as we venture into the realm of nanometer geometry semiconductors and their failure modes. The uncertainty about the future of long-established methods has been many in the industry seeking alternative methods.

At the same time, on the component supplier side, semiconductor suppliers have been able to provide such substantial increases in component reliability and operational lifetimes that they are slowly beginning to drop the military standard MIL-STD-883B testing methods, and many have dropped their lines of military specification (mil-spec) parts. A major reason contributing to this is that instead of focusing on mil-spec parts, they have moved their focus to commercial-grade parts, where the unit volumes are much higher. In recent times the purchasing power of military markets has diminished to the point where they no longer have the dominant presence and leverage. Instead, system builders took their commercial-grade devices, sent them out to testing labs, and found that the majority of them would, in fact, operate reliably at the extended temperature ranges and environmental conditions required by the mil-spec. In addition, field data gathered over the years has improved much of the empirical data necessary for complex algorithms for reliability prediction [10–14].

IN 1994, A MEMORANDUM BY THE THEN SECRETARY OF DEFENSE, WILLIAM J. PERRY INSTRUCTED THE DEPARTMENT OF DEFENSE TO USE COMMERCIAL DEVELOPMENT METHODS TO THE MAXIMUM EXTENT POSSIBLE WHEN ACQUIRING NEW WEAPONS. THIS ESSENTIALLY DID AWAY WITH MIL-SPEC METHODS OF DEVELOPMENT REQUIREMENTS AND INSTEAD CREATING REQUIREMENTS BASED ON OBJECTIVE OUTCOMES. COMMERCIAL CONTRACTORS WERE THEN ALLOWED TO USE THEIR OWN METHODS TO ACHIEVE THOSE GOALS. THIS POLICY CHANGE RECOGNIZED THE ADVANCES OF COMMERCIAL ENGINEERING, FAR EXCEEDED THOSE OF THE OLD

MILITARY SPECIFICATIONS, AND SHOULD BE USED WHEREVER POSSIBLE. HENCE, WHAT YOU SAID ABOVE IS TRUE, BUT DEFINITELY IT WAS ENHANCED BY THIS IMPORTANT POLICY CHANGE WITHIN THE DOD. FOR MORE INFORMATION, GO TO HTTPS://WWW.SAE.ORG/STANDARDSDEV/MILITARY/MILPERRY.HTM

The European Power Supply Manufacturers Association [15] provides engineers, operations managers, and applied statisticians with both qualitative and quantitative tools for solving a variety of complex, real-world reliability problems. There is wealth of accompanying examples and case studies as follows [15]:

- Comprehensive coverage of assessment, prediction, and improvement at each stage of a product's life cycle.
- Clear explanations of modeling and analysis for hardware ranging from a single part to whole systems.
- Thorough coverage of test design and statistical analysis of reliability data.
- The software reliability.
- Coverage of effective management of reliability, product support, testing, pricing, and related topics.
- Lists of sources for technical information, data, and computer programs.
- Hundreds of graphs, charts, and tables, as well as over 500 references.

There are very seldom efficiency prediction references, because this area is undeveloped. But there are many references in reliability prediction. For example, here are some references in this area. For example, Gipper [16] provides a comprehensive overview of both qualitative and quantitative aspects of reliability. Mathematical and statistical concepts related to reliability modeling and analysis are presented along with a necessary bibliography and a listing of resources, including journals, reliability standards, other publications, and databases. The coverage of individual topics is not always deep but should provide a valuable reference for engineers or statistical professionals working in the reliability testing field.

There are many other publications (mostly articles and papers) that relate to the current situation in the methodological aspects of reliability prediction.

In the Reliability and Maintainability Symposium (RAMS) Proceedings alone, there have been more than 100 papers published in this area. For example, RAMS 2012 published six papers. Most of them related to reliability prediction methods in software design and development.

Both physics-based modeling and simulation and empirical reliability have been subjects of much interest in computer graphics.

The following provides the basic content of the abstracts from some of the articles in reliability prediction from the RAMS Proceedings:

1. Cai et al. [17] present a novel method of field reliability prediction considering environment variation and product individual dispersion. Wiener diffusion process with drift was used for degradation modeling, and a link function which presents degradation rate is introduced to model the impact of varied environments and individual dispersion. Gamma, transformed-Gamma (T-Gamma), and

the normal distribution with different parameters are employed to model right-skewed, left-skewed, and symmetric stress distribution in the study case. Results show obvious differences in reliability, failure intensity, and failure rate compared with a constant stress situation and with each other. The authors indicate that properly modeled (proper distribution type and parameters) environmental stress is useful for varied environment-oriented reliability prediction.

2. Chigurupati et al. [18] explore the predictive abilities of a machine learning technique to improve upon the ability to predict individual component times until failure in advance of actual failure. Once failure is predicted, an impending problem can be fixed before it actually occurs. The developed algorithm was able to monitor the health of 14 hardware samples and notify us of an impending failure providing time to fix the problem before actual failure occurred.

3. Wang et al. [19, 20] deal with the concept of space radiation environment reliability for satellites and establish a model of space radiation environment reliability prediction, which establishes the relationship among system failure rate and space radiation environment failure rate and non-space radiation environment failure rate. It provides a method of space radiation environment reliability prediction from three aspects:

   (a) Incorporating the space radiation environment reliability into traditional reliability prediction methods, such as FIDES and MIL-HDBK-217.
   (b) Summing up the total space radiation environment reliability failure rate by summing the total hard failure rate and soft failure rate of the independent failure rates of SEE, total ionizing dose (TID), and displacement damage (DD).
   (c) Transferring TID/DD effects into equivalent failure rate and considering single event effects by failure mechanism in the operational conditions of duty hours within calendar year. A prediction application case study has been illustrated for a small payload. The models and methods of space radiation environment reliability prediction are used with ground test data of TID and single event effects for field programmable gate arrays.

4. In order to utilize the degradation data from hierarchical structure appropriately, Wang et al. first collected and classified feasible degradation data from a system and subsystems [20]. Second, they introduced the support vector machine method to model the relationship among hierarchical degradation data, and then all degradation data from subsystems are integrated and transformed to the system degradation data. Third, with this processed information, a prediction method based on Bayesian theory was proposed, and the hierarchical product's lifetime was obtained. Finally, an energy system was taken as an example to explain and verify the method in this paper; the method is also suitable for other products.

5. Jakob et al. used knowledge about the occurrence of failures and knowledge of the reliability in different design stages [21]. To show the potential of this approach, the paper presents an application for an electronic braking system. In a first step, the approach presented shows investigations of the physics of failure

based on the corresponding acceleration model. With knowledge of the acceleration factors, the reliability can be determined for each component of the system for each design stage. If the failure mechanisms occurring are the same for each design stage, the determined reliability of earlier design stages could be used as preknowledge as well.

Because that is difficult in many cases, and further investigations are deemed necessary.

6. Today's complex designs, which have intricate interfaces and boundaries, cannot rely on the MIL-HDBK-217 methods to predict reliability. Kanapady and Adib [22] present a reliability prediction approach for design and development of projects that demand high reliability where the traditional prediction approach has failed to do the job. The reliability of a solder ball was predicted. Sensitivity analysis, which determines factors that can mitigate or eliminate the failure mode(s), was performed. Probabilistic analysis, such as the burden capability method, was employed to assess the probability of failure mode occurrences, which provides a structured approach to ranking of the failure modes, based on a combination of their probability of occurrence and severity of their effects.

7. Microelectronics device reliability has been improving with every generation of technology, whereas the density of the circuits continues to double approximately every 18 months. Hava et al. [23] studied field data gathered from a large fleet of mobile communications products that were deployed over a period of 8 years in order to examine the reliability trend in the field. They extrapolated the expected failure rate for a series of microprocessors and found a significant trend, whereby the circuit failure rate increases approximately half the rate of the technology, going up by approximately $\sqrt{2}$ in that same 18-month period.

8. Thaduri et al. [24] studied the introduction, functioning, and importance of a constant fraction discriminator in the nuclear field. Furthermore, the reliability and degradation mechanisms that affect the performance of the output pulse with temperature and dose rates act as input characteristics as properly explained. Accelerated testing was carried out to define the life testing of the component with respect to degradation in output transistor-transistor logic pulse amplitude. Time to failure was to be properly quantified and modeled accordingly.

9. Thaduri et al. [25] also discussed several reliability prediction models for electronic components, and comparison of these methods was also illustrated. A combined methodology for comparing the cost incurred for prediction was designed and implemented with an instrumentation amplifier and a bipolar junction transistor (BJT). By using the physics of failure approach, the dominant stress parameters were selected on the basis of a research study and were subjected to both an instrumentation amplifier and a BJT. The procedure was implemented using the methodology specified in this paper and modeled the performance parameters accordingly.

From the prescribed failure criteria, the MTTF was calculated for both the components. Similarly, using the 217Plus reliability prediction book, the MTTF was also calculated and compared with the prediction using physics of failure. Then, the costing implications of both the components were discussed and compared. For critical components like an instrumentation amplifier, it was concluded that though the initial cost of physics of failure prediction is too high, the total cost incurred, including the penalty costs, is lower than that of a traditional reliability prediction method. But for noncritical components like a BJT, the total cost of physics of failure approach was too high compared with a traditional approach, and hence a traditional approach was more efficient. Several other factors were also compared for both reliability prediction methods.

Much more literature on methodological approaches to reliability prediction are available. Here are some good examples.

The purpose of the MIL-HDBK-217F handbook [26] is to establish and maintain consistent and uniform methods for estimating the inherent reliability (i.e., the reliability of a mature design) of military electronic equipment and systems. It provides a common basis for reliability predictions during acquisition programs for military electronic systems and equipment. It also establishes a common basis for comparing and evaluating reliability predictions of related or competitive designs.

Another document worthy for discussion is the Telcordia document, Issue 4 of SR-332 [27].

This provides all the tools needed for predicting device and unit hardware reliability and contains important revisions to the document.

The *Telcordia Reliability Prediction Procedure* has a long and distinguished history of use both within and outside the telecommunications industry. Issue 4 of SR-332 provides the only hardware reliability prediction procedure developed from the input and participation of a cross section of major industrial companies. This lends the procedure and the predictions derived from it a high level of credibility, free from the bias of any individual supplier or service provider.

Issue 4 of SR-332 contains the following:

- Recommended methods for prediction of device and unit hardware reliability. These techniques estimate the mean failure rate in FITs for electronic equipment. This procedure also documents a recommended method for predicting serial system hardware reliability.
- Tables needed to facilitate the calculation of reliability predictions.
- Revised generic device failure rates, based mainly on new data for many components.
- An extended range of complexity for devices, and the addition of new devices.
- Revised environmental factors based on field data and experience.

Section Summary
1. Most of the approaches of this section considered are difficult to use successfully in practice for efficiency prediction.
2. The basic problem for this is a lack of close connection to the source, which is necessary for obtaining accurate initial information that is needed for calculating changing efficiency components and their parameters during the product's life cycle.

3. Since the efficiency prediction methodology was not developed, the following basic methods of more developed traditional reliability prediction were discussed:

   - Empirical reliability prediction methods that are based on the statistical analysis of historical failure data models, developed from statistical curves. These methods are not considered accurate simulations of field situations and are an obstacle to obtaining accurate initial information for calculating the dynamics of changing failure (degradation) parameters during a product or technology service life.
   - Physics of failure approach, which is based on the understanding of the failure mechanism and applying the physics of failure model to the data.

   However, they are not considered approaches for obtaining data for service life during the design and manufacturing stages of a new model of product/technology. Accurate initial information from the field during service life is not available during these stages of development.

   - Laboratory-based or proving ground-based life testing reliability prediction, which uses accelerated life testing (ALT) in the laboratory, but does not accurately simulate the real-world conditions, changing interacted parameters encountered in the field during service life of the product/technology.

   As it will be shown later in this book, reliability is only one from many interconnected components of efficiency. Therefore, considering it separately from other components leads to unsuccessful efficiency prediction. This is the basic negative aspect of current technologies of reliability prediction.

4. Recalls, complaints, injuries and deaths, and significant costs are direct results of these prediction failures.
5. Real products rarely exhibit a constant failure rate and, therefore, cannot be accurately described by exponential, log-normal, or other traditionally used theoretical distributions. Real-life failure rates are mostly random.
6. Reliability prediction is often considered as a separate issue, but, in real life, reliability is only one essential interactive element of a product/technology efficiency prediction [30].

Analysis has shown that the current status of product/technology methodological aspects of reliability prediction for industries such as electronic, automotive, aircraft, aerospace, off-highway, farm machinery, and others are not very successful. The basic cause is the difficulty of obtaining accurate initial information for specific product prediction calculations during the real-world use of the product.

Accurate prediction requires information similar to that experienced in the real world.

There are many other methodological aspects of reliability, as well as efficiency prediction that commonly have similar problems. For example, see Refs. [31–40] in the References below.

2.2　Why the Level of New Product Current Prediction Is Low

From the review in the Sect. 2.1, there are problems that can be readily seen as follows:

- How can one obtain common methodological (strategic) aspects of products and interacted components successful efficiency prediction?
- As one can see from Fig. 2.1, reliability is one from many interacted components of the product's efficiency. Therefore, one needs to understand that if we consider reliability separately, it is different from that in real-life situation reliability. Therefore, reliability prediction is only one from interacted components of product's efficiency in real life. One can read in [44] the full product efficiency prediction consideration.
- How can one obtain accurate initial information for each particular product successful efficiency prediction, including quality, safety, reliability, durability, profit, life cycle cost, and others?

For solution of the above problems, one needs to understand the basic reasons why presently used approaches cannot help to solve these problems.

About Other Methodological Aspects
As was mentioned above, there are many publications in reliability prediction methodology. For example, RAMS 2012 Proceedings published six papers in this area. Most of them relate to theoretical methods in software design and development. Both physics-based modeling and simulation and empirical reliability have received interest in computer graphics.

A brief description of the above issues follows:

Usually, one provides reliability and durability evaluation directly after stress testing (in laboratory or proving grounds). This evaluation relates to test conditions, i.e., laboratory or proving ground conditions, but not to field conditions. If one wants to know the test subject's reliability or durability in the field, one can only use the above testing results for prediction of reliability and durability in the field. For this purpose, one must use prediction methods which are normally have at best a probabilistic character. This is a more complicated way than directly evaluation after testing, and therefore, one seldom uses this method. As a result, one cannot use the above evaluation for accurately predicting the product reliability and durability during any given time.

Usually, one uses separate vibration testing or temperature/humidity testing or testing with simulation several types of field input influences. But in the real world, many input influences act on the test subject. They are temperature, humidity, air pollution (mechanical and chemical), radiation (visible, ultraviolet, and infrared), air fluctuation, features of the road (type of road, profile, density, etc.), speed of movement, input voltage, electrostatic discharge, and others. If one uses only simulating several from the above multi-environmental real-world input

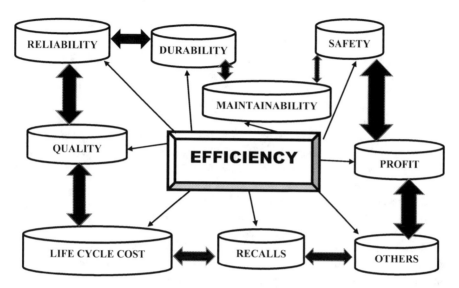

Fig. 2.1 Efficiency as a complex of interacted components

influences and ignores their interconnections, which are influencing reliability and durability and, finally, efficiency, inaccurate results will occur.

There are some of the reasons why current types of accelerated stress testing cannot help to predict accurately the reliability, durability, and other efficiency components:

1. Often one uses only components (details) and assemblies (units) stress testing, but not stress testing an entire product. In this case, one is ignoring the interconnections between components and assemblies. But, the stress testing results are different from usage of a test subject in the real world. Therefore, using these results cannot predict accurately the reliability, durability, and other efficiency components of the product in the field.
2. Inaccurate simulation and mistakes in using stress testing leads to inaccurate prediction of product quality, reliability, durability, and other components. This is one from the basic reasons of recalls.

   Therefore, actual profits are often several times lower than their potentials. See Fig. 2.2 below as a typical example.

   As was written by Misra [12], "Money was saved years earlier by gambling with a substandard reliability program," but as shown in Fig. 2.2, the short-term gain was not a good long-term investment.
3. Often industrial companies have used accelerated stress testing with loads more than maximum field loads. This changed the physics of degradation (or chemistry of degradation) process during this testing in comparison with the field physics of degradation (chemistry of degradation) process. As a result, the time to

failures, as well as the number of failures during this testing, can be different from the failures in the field situation.

4. One has to take into account that each failure mechanism responds to stress differently, and each component of the product has different failure mechanism. Using accelerated test data plus a single acceleration factor can result in a MTBF estimate which is erroneous and misleading [13].

5. Often one forget that there is no standard stress stimulus portfolio—each product program will have differences, because of the vast diversity of products [13].

6. Components of efficiency demonstration based on statistically determined sample sizes are often invalid, because the samples would not be a true random representation of the production parts [11].

7. The acceptation process for the field loading simulation for stress testing is not based on a true representative region, which represents all areas (including climatic) of the product use in the field.

8. Specifically, one often finds the usage of accelerated testing with simulation of constant stress, or step stress, or cycling stress. This contradicts the real-world situation, where the loading has a random stress character.

The problems described above can be practically eliminated if one will use accelerated reliability and durability testing technology (ART/ADT). Description of these types of testing can be read briefly in Chap. 4 and in publications [30, 44, 60–63]. The basic purpose of the above publications is to show how one can simulate accurately field situation for obtaining initial information for accurate efficiency development and prediction.

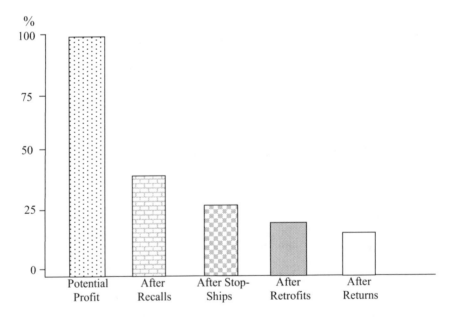

Fig. 2.2 Effect of poor efficiency prediction on company profits

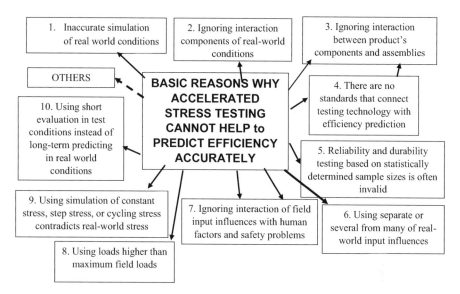

Fig. 2.3 The basic reasons why accelerated stress testing cannot help to predict efficiency accurately

For successful providing ART/ADT technology, a multidisciplinary team will be needed to engineer and manage the application of this technology to a particular product. The team should include, as a minimum, the following (Fig. 2.3):

- A team leader who is a high-level manager who understands the strategy of providing this technology, the principles of accurate simulation of the field situation, and knows what professional disciplines need to be included on the team.
- A program manager to act as a guide through the process and remove any barriers that prevent the team from succeeding and has a knowledge in design and technology for meeting the test objectives.
- Engineering resources to perform unit filtering (selection and elimination), failure analysis, chemical problems solution in simulation, physical problems solution in simulation, prediction methodology, system of control development, design, diagnostic and corrective action for mechanical, electrical, hydraulic, etc., as well as both hardware and software development and implementation.

The team must work in close contact with departments responsible for design, manufacturing, marketing, and selling. One can see the explanation of this concept in [60].

There are many practical examples showing how progress with successful prediction and accelerated reliability/durability testing as one of the basic components of this prediction is moving forward very slowly (see Chap. 5).

2.3   Example of Low-Level Efficiency Prediction

Aerospace is one from more complicated and developed areas of design engineering. Therefore, the following example demonstrates how level of testing and efficiency prediction development are much lower than product design and manufacturing development. As a result, many financial losses have occurred.

Let us show this from an example of the NASA Rover mission to Mars and its landing platform. The first Rover on Mars and its landing platform Mariner 4 was developed by NASA and sent to Mars. This vehicle was sent to land photos of the Mars surface. A developed vehicle was sent in 1998. The author was familiar with it before it was sent to Mars.

In 1997 the author had presented a paper on Accelerated Evaluation of Agricultural Equipment Reliability at the ASAE Annual International Meeting in Kansas City. After this presentation he was invited by professor Joseph Zayaz to visit Kansas State University. The professor introduced Dr. Klyatis to the engineering departments of this university. In one department Dr. Klyatis saw the research center that worked for NASA. He saw in this center the Rover being tested that NASA planned on sending to Mars. The level of testing was so poor as the author has described in this book on current testing analysis.

In 1998 the author made a presentation (Fig. 2.4) at a IEEE 1998 Workshop on Accelerated Stress Testing (September 1998, Pasadena, California), published in IEEE 1998 Accelerated Stress Testing Workshop Proceedings. During this workshop, the headquarters (see Fig. 2.5) invited Dr. Klyatis to visit the NASA Jet Propulsion Laboratory (JPL) with them.

During the visit JPL professionals of NASA showed the group the research station for rover. As the author was familiar with this Rover testing program at the Kansas State University research center, he asked the JPL professionals how long they were planning to use this machine for gathering and receiving information from Mars to Earth. The answer was: "We guarantee 90 days of working this station at the Mars." The author answered them: "You are wrong. The rover will not work on Mars." The JPL professional only smiled.

Several weeks later the Rover was sent to Mars. Afterwards, the author read in *The New York Times* that the Rover died after 3 days, which was nowhere near the warranty standard of 90 days that NASA gave. After that sad result, JPL employees called the author and asked: "Why did you say that the Rover will not work?" His answer was: "Your testing technology is poor. You cannot accurately simulate the real Mars conditions for conducting a good test. Therefore, you cannot accurately predict the product's reliability, durability, and other efficiency components. Therefore, the warranty period was wrong." As a result, the author was sure this Rover will not work at the expected Mars conditions.

One detail example: first, the daily temperature range on Mars goes from approximately +100 °C to −100 °C. And second, the speed of this change before night increases much faster than during the day. During the simulation of Mars' environment, the researchers did not take these variables into account. As a result, because of this poor simulation and prediction, the USA spent over hundreds of millions of dollars without positive results.

Fig. 2.4 Dr. Lev Klyatis's presentation at the IEEE Reliability Society Workshop, Pasadena, CA

Fig. 2.5 Paul Parker (IEEE workshop chair of technical program), Dr. Lev Klyatis, and this work-shop registration Chair Kirk Gray during IEEE 1998 Workshop on Accelerated Stress Testing, Pasadena, California

Several years later, the author visited Langley Research Center, NASA with experts from the SAE G-11 Division in development of standardization in aerospace (see Fig. 4.12 in [1]).

The employees of this center could not answer on author's question: "Who is working here in improvement methodology of reliability/durability testing?" After that and visiting Lockheed Martin and other advanced companies of the USA and Germany, it was understandable how the level of expertise at the TESTMASH company was higher (See Chap. 6). When the author learned from other sources about how consistently unsuccessful is product efficiency at the SAE G-11 Division, he developed six SAE standards with common title "Reliability Testing."

The drafts of these standards were included in his books [1, 11, 30] that were published by Elsevier, Wiley, and SAE International.

References

1. Lev Klyatis, Edward Anderson (2018) Reliability Prediction and Testing. John Wiley & Sons.
2. Naval Surface Warfare Center, Carderock Division (2011) *Handbook of Reliability Prediction Procedures for Mechanical Equipment*. Logistic Technology Support, Carderockdiv, NSWC-10. West Naval Surface Warfare Center, Carderock Division, Bethesda, MD.
3. Atiyeh C, Blackwell R. (2016) Massive Takata airbag recall: everything you need to know, including full list of affected vehicles. Update: December 29. *Car and Driver*. https://blog.caranddriver.com/massivetakata-airbag-recall-everything-you-need-to-know-including-fulllist-of-affected-vehicles/ (accessed January 18, 2018).
4. SAE International (2016) UPDATE 12/29/2016.
5. SR-332, Issue 2 (2006) Reliability Prediction Procedure for Electronic Equipment, Telcordia, September
6. Web Center for Social Research Methods. (2006). Types of reliability, in *The Research Methods Knowledge Base*. https://www.socialresearchmethods.net/kb/reltypes.php (accessed January 18, 2018).
7. Weibull.com (2005). Standards based reliability prediction: applicability and usage to augment RBDs. Part I: introduction to standards based reliability prediction and lambda predict. *Reliability HotWire* Issue 50 (April). http://www.weibull.com/hotwire/issue50/hottopics50.htm (accessed January 18, 2018).
8. ReliaSoft (2018) Reliability prediction methods for electronic products. *Reliability EDGE* **9**(1). http://www.reliasoft.com/pubs/reliabilityedge_v9i1.pdf (accessed January 18, 2018).
9. ANSI/VITA 51.2 (2016). *Physics of Failure Reliability Predictions*.
10. Blischke WR, Prabhakar Murthy DN. (2000). *Reliability: Modeling, Prediction, and Optimization*. John Wiley & Sons.
11. Lev M. Klyatis (2012) Accelerated Reliability and Durability Testing Technology. John Wiley & Sons
12. Misra KB. (1992). *Reliability Analysis and Prediction, Volume 15. A Methodology Oriented Treatment*. Elsevier Science.
13. Jones TL (2010). *Handbook of Reliability Prediction Procedures for Mechanical Equipment*. Naval Surface Warfare Center, Carderock Division.
14. An YH, Draughn RA. (1999). *Mechanical Testing of Bone–Implant Interface*. CRC Press.
15. European Power Supply Manufacturers Association. (2005). *Guidelines to understanding reliability prediction*. http://www.epsma.org/MTBF%20Report_24%20June%202005.pdf (accessed January 18, 2018).
16. Gipper J. (2014). *Choice of Reliability Prediction Methods*. http://vita.mil-embedded.com/articles/choice-reliability-prediction-methods/(accessed February 2, 2018).
17. Cai Y-K, Wei D, Ma X-B, Zhao Y (2015) Reliability prediction method with field environment variation. In *2015 Annual Reliability and Maintainability Symposium (RAMS)*. IEEE Press, pp. 1–7.
18. Chigurupati A, Thibaux R, Lassar N. (2016) Predicting hardware failure using machine learning. In *2016 Annual Reliability and Maintainability Symposium (RAMS)*. IEEE Press, pp. 1–6.

19. Wang Q, Chen D, Bai H. (2016). A method of space radiation environment reliability prediction. In *2016 Annual Reliability and Maintainability Symposium (RAMS)*. IEEE Press, pp. 1–6.
20. Wang L, Zhao X, Wang X, Mu M. (2016). A lifetime prediction method with hierarchical degradation data. In *2016 Annual Reliability and Maintainability Symposium (RAMS)*. IEEE Press, pp. 1–6.
21. Jakob F, Schweizer V, Bertsche B, Dobry A. (2014). Comprehensive approach for the reliability prediction of complex systems. In *2014 Reliability and Maintainability Symposium*. IEEE Press, pp. 1–6.
22. Kanapady R, Adib R. (2013) Superior reliability prediction in design and development phase. In *2013 Proceedings Annual Reliability and Maintainability Symposium (RAMS)*. IEEE Press, pp. 1–6.
23. Hava A, Qin J, Bernstein JB, Bo Y. (2013) Integrated circuit reliability prediction based on physics-of-failure models in conjunction with field study. In *2013 Proceedings Annual Reliability and Maintainability Symposium (RAMS)*. IEEE Press, pp. 1–6.
24. Thaduri A, Verma AK, Gopika V, Kumar U. (2013) Reliability prediction of constant fraction discriminator using modified PoF approach. In *2013 Proceedings Annual Reliability and Maintainability Symposium (RAMS)*. IEEE Press, pp. 1–7.
25. Thaduri A, Verma AK, Kumar U. (2013) Comparison of reliability prediction methods using life cycle cost analysis. In *2013 Proceedings Annual Reliability and Maintainability Symposium (RAMS)*. IEEE Press, pp. 1–7.
26. DAU. (1995). *MIL-HDBK-217F (Notice 2). Military Handbook: Reliability Prediction of Electronic Equipment*. Department of Defense, Washington, DC.
27. Telcordia. (2016). *SR-332, Issue 4, Reliability Prediction Procedure for Electronic Equipment*.
28. Lu H, Kolarik WJ, Lu SS. (2001). Real-time performance reliability prediction. *IEEE Transactions on Reliability* 50(4): 353–357.
29. NIST/SEMATECH. (2010). Assessing product reliability. In *Engineering Statistics e-Handbook*. US Department of Commerce, Washington, DC, chapter 8.
30. Klyatis L. (2016) *Successful Prediction of Product Performance: Quality, Reliability, Durability, Safety, Maintainability, Life Cycle Cost, Profit, and Other Components*. SAE International, Warrendale, PA.
31. DAU. (1991). *MIL-HDBK-217F (Notice 1). Military Handbook: Reliability Prediction of Electronic Equipment*. Department of Defense, Washington, DC.
32. Telcordia. (2001). *SR-332, Issue 1, Reliability Prediction Procedure for Electronic Equipment*.
33. Telcordia. (2006). *SR-332, Issue 2, Reliability Prediction Procedure for Electronic Equipment*.
34. ITEM Software and ReliaSoft Corporation. (2015). *D490 Course Notes: Introduction to Standards Based Reliability Prediction and Lambda Predict*.
35. Foucher B, Boullie J, Meslet B, Das D. (2002). A review of reliability prediction methods for electronic devices. *Microelectronics Reliability* 42(8): 1155–1162.
36. Pecht M, Das D, Ramakrishnan A. (2002). The IEEE standards on reliability program and reliability prediction methods for electronic equipment. *Microelectronics Reliability* 42: 1259–1266.
37. Talmor M, Arueti S. (1997). Reliability prediction: the turnover point. In *Annual Reliability and Maintainability Symposium: 1997 Proceedings*. IEEE Press, pp. 254–262.
38. Hirschmann D, Tissen D, Schroder S, de Doncker RW (2007). Reliability prediction for inverters in hybrid electrical vehicles. *IEEE Transactions on Power Electronics*, 22(6): 2511–2517.
39. NIST Information Technology Library. https://www.itl.nist.gov.
40. SeMaTech International. (2000) *Semiconductor Device Reliability Failure Models*. www.sematech.org/docubase/document/3955axfr.pdf (accessed January 19, 2018).
41. Nicholls D (2012). An objective look at predictions – ask questions, challenge answers. In *2012 Proceedings Annual Reliability and Maintainability Symposium*. IEEE Press, pp. 1–6.
42. Theil N (2016). Fatigue life prediction method for the practical engineering use taking in account the effect of the overload blocks. *International Journal of Fatigue* 90: 23–35.
43. Denson W (1998). The history of reliability prediction. *IEEE Transactions on Reliability* 47(3-SP, Part 2): SP-321–SP-328.

44. Klyatis L (2014). The role of accurate simulation of real-world conditions and ART/ADT technology for accurate efficiency predicting of the product/process. *SAE 2014 World Congress*, paper # 2014-01-0746.
45. Jones JA (2008). Electronic reliability prediction: a study over 25 years. PhD thesis, University of Warwick.
46. Wong KL (1990). What is wrong with the existing reliability prediction methods? *Quality and Reliability Engineering International* **6**(4): 251–257.
47. Black AI (1989). Bellcore system hardware reliability prediction. In *Proceedings Annual Reliability and Maintainability Symposium*.
48. Bowles JB (1992). A survey of reliability prediction procedures for microelectronic devices. *IEEE Transactions in Reliability* **41**: 2–12.
49. Chan HT, Healy JD (1985). Bellcore reliability prediction. In *Proceedings Annual Reliability and Maintainability Symposium*.
50. Healy JD, Aridaman KJ, Bennet JM (1999). Reliability prediction. In *Proceedings Annual Reliability and Maintainability Symposium*.
51. Leonard CT, Recht M (1990). How failure prediction methodology affects electronic equipment design. *Quality and Reliability Engineering International* **6**: 243–249.
52. Wymysłowski A (2011). Editorial. 2010 EuroSimE international conference on thermal, mechanical and multi-physics simulation and experiments in micro-electronics and micro-systems. *Microelectronics Reliability* **51**: 1024–1025
53. Kulkarni C, Biswas G, Koutsoukos X (2010). Physics of failure models for capacitor degradation in DC–DC converters. https://c3.nasa.gov/dashlink/static/media/publication/2010_MARCON_DCDCConverter.pdf (accessed January 19, 2018).
54. Eaton DH, Durrant N, Huber SJ, Blish R, Lycoudes N (2000). *Knowledge-based reliability qualification testing of silicon devices*. International SEMATECH Technology Transfer # 00053958A-XFR. http://www.sematech.org/docubase/document/3958axfr.pdf (accessed January 19, 2018).
55. Osterwald CR, McMahon TJ, del Cueto JA, Adelstein J, Pruett J (2003). Accelerated stress testing of thin-film modules with SnO2:F transparent conductors. Presented at the *National Center for Photovoltaics and Solar Program Review Meeting Denver,* Colorado. https://www.nrel.gov/docs/fy03osti/33567.pdf (accessed January 19, 2018).
56. Vassiliou P, Mettas A (2003). Understanding accelerated life-testing analysis. In *2003 Annual Reliability and Maintainability Symposium*.
57. Mettas A (2010). Modeling and analysis for multiple stress-type accelerated life data. In *46th Reliability and Maintainability Symposium*.
58. Dodson B, Schwab H (2006). *Accelerated Testing: A Practitioner's Guide to Accelerated and Reliability Testing*. SAE International, Warrendale, PA.
59. Ireson WG, Combs CF Jr, Moss RY (1996). *Handbook on Reliability Engineering and Management*. McGraw-Hill.
60. Klyatis LM, Klyatis EL (2006). *Accelerated Quality and Reliability Solutions*. Elsevier.
61. Klyatis LM, Klyatis EL (2002). *Successful Accelerated Testing*. Mir Collection, New York.
62. Klyatis LM, Verbitsky D (2010) Accelerated Reliability/Durability Testing as a Key Factor for Accelerated Development and Improvement of Product/Process Reliability, Durability, and Maintainability. SAE Paper 2010-01-0203. Detroit. 04/12/2010. (Also in the book SP-2272).
63. Klyatis L (2009). Specifics of accelerated reliability testing. In *IEEE Workshop Accelerated Stress Testing. Reliability (ASTR 2009)* [CD], October 7–9, Jersey City.
64. Klyatis L, Vaysman A (2007/2008). Accurate simulation of human factors and reliability, maintainability, and supportability solutions. *The Journal of Reliability, Maintainability, Supportability in Systems Engineering* (Winter).
65. Klyatis L. (2006). Elimination of the basic reasons for inaccurate RMS predictions. In *A Governmental–Industry Conference "RMS in A Systems Engineering Environment"*, DAU-West, San Diego, CA, October 11–12.

Chapter 3
Technology of Successful Prediction of New Product Efficiency (Quality, Reliability, Durability, Maintainability, Safety, Life Cycle Cost, Profit, and Other Components)

3.1 The Basis of Successful Prediction of Product Efficiency

For understanding the prediction system, one must comprehend the interconnection relationship of design, manufacturing, and usage. They are related as shown on Fig. 3.1. They are grouped, because each of the three components is in itself a complex of procedures. For example, the design group consists of multiple research projects, investigations, calculations, and different levels and types of testing and works with suppliers, ordered companies, or organizations.

All of them must be taken into account and analyzed for a successful prediction of product efficiency.

The basis of a successful prediction is connected with an accurate simulation of real-world conditions. One can see in Fig. 3.2 schematically the three basic interconnected complexes of real-world conditions.

Real input influences of field conditions depend on type of studied equipment, climatic conditions of usage, and some subjective factors. For example, one can see in Fig. 3.3 the following interconnected some input influences of the field conditions for automobiles.

Figure 3.4 demonstrates an example of human factors that influence on the product/process efficiency.

The human factors are not separate factors. They interact with other basic components of real-world conditions.

Safety factors are the combination of two basic components: risk problems and hazard analysis.

The two basic safety components are a combination of various sub-components. The solution of the risk problem is found in the following sub-components:

Supplementary Information The online version contains supplementary material available at https://doi.org/10.1007/978-3-031-16655-6_3.

L. M. Klyatis, *Prediction Technologies for Improving Engineering Product Efficiency*, https://doi.org/10.1007/978-3-031-16655-6_3

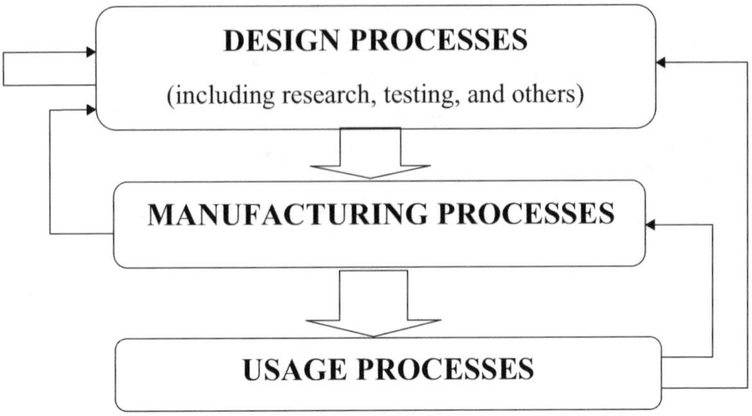

DESIGN PROCESSES

(including research, testing, and others)

MANUFACTURING PROCESSES

USAGE PROCESSES

Fig. 3.1 The feedback between processes of machines life cycle

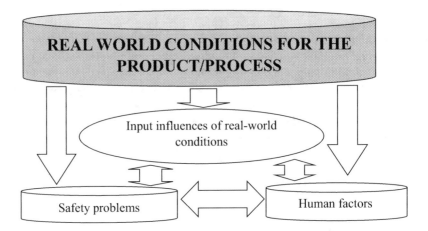

REAL WORLD CONDITIONS FOR THE PRODUCT/PROCESS

Input influences of real-world conditions

Safety problems

Human factors

Fig. 3.2 Three basic interconnected complexes of real-world conditions of the product/process

- Risk assessment.
- Risk management.
- Risk evaluation.

Each of these sub-components consists of sub-sub-components. For example, risk assessment consists of the following five sub-sub-components, as shown in Fig. 2.8 [1]:

- Risk estimation
- Risk analysis
- Risk evaluation
- Risk reduction
- Hazard identification

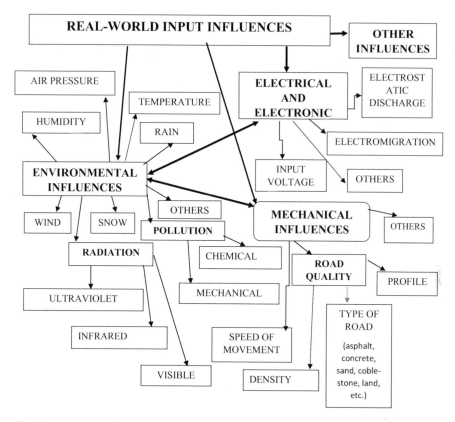

Fig. 3.3 Scheme of different real-world input influences for automobiles

Solving the safety problem requires a simultaneous study and evaluation of the full complex of these interacting components and sub-components. This will provide an accurate prediction for accelerated development and improvement of the product/process.

To obtain information for risk assessment, one needs to know the following:

- Limits of the machinery's capabilities
- Accident and incident history
- Requirements for the life phases of the machinery
- Basic design drawing that demonstrate the nature of the machinery
- Statements about damage to health

For risk analysis one needs:

- Identification of hazards
- Methods of setting limits for the machinery
- Risk estimation

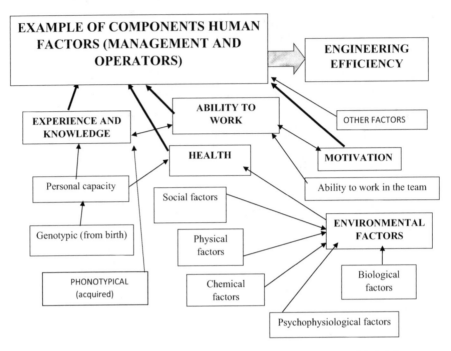

Fig. 3.4 Basic components of human factors (management and operators) and their influence on engineering efficiency

For risk reduction, it is assumed that the user has a role to play in the risks being seen.

The basis of successful prediction of product efficiency includes an accurate real-world simulation. One can see from Fig. 3.5 the example what the path from real-world to efficiency means if we take into account one from three components of this real-world input influence.

3.2   The Strategy of Successful Efficiency Prediction

One can see the common scheme of successful efficiency prediction strategy in Fig. 3.6.

The strategy of successful efficiency prediction consists of five common steps (Fig. 3.7):

- Accurate simulation of the real-world conditions
- Accelerated reliability/durability testing (ART/ADT)
- Methodology of efficiency prediction
- Successful efficiency prediction
- Improving engineering product efficiency

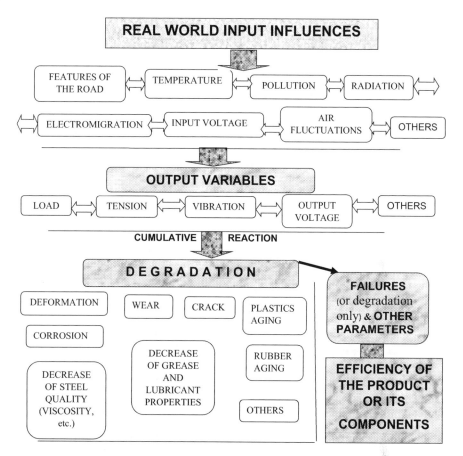

Fig. 3.5 The path from real-world input influences on the product to efficiency of the product or its components (including quality, reliability, durability, maintainability, safety, life cycle cost, profit, and others)

The author created his ideas of successful efficiency prediction while working in the USSR, where he created the strategy and components, methodology, and tools for this new scientific engineering direction. On the basis of this work with farm machinery, he wrote his dissertation "The Basis of Successful Prediction of Farm Machinery" and earned his doctorate of technical science. Then, he continued his work even after moving to the United States, where he is still improving these concepts and expanding their use in different areas of automotive and other areas of engineering.

As was shown in Fig. 3.6, this strategy includes two basic components, the details of which were published in English in the books [1–5], in Chinese in the book [7], and in Russian in [8]. The strategy described above is related to the following scheme of successful prediction of product efficiency (Fig. 3.8).

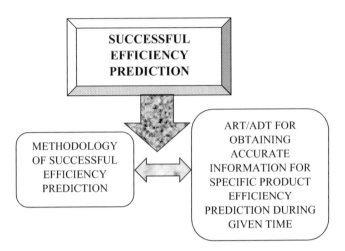

Fig. 3.6 Common strategy of successful efficiency prediction (two basic components)

Fig. 3.7 Five common steps for improving engineering product efficiency through technologies of successful efficiency prediction

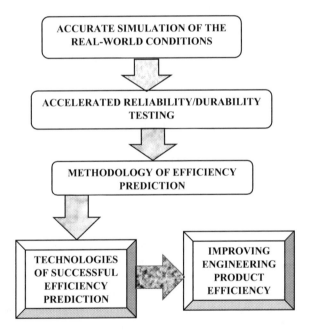

3.3   Methodological Aspects (The First Key Factor) of Successful Efficiency Prediction

One can study the full methodology in author's cited books, published by Elsevier, Wiley, and others [1–8], as well as publications [16–20] and others. Below can be seen some aspects of this methodology that includes (Fig. 3.9):

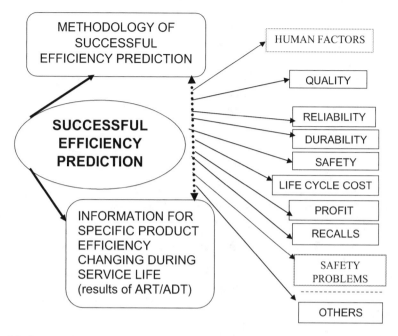

Fig. 3.8 Strategical scheme for successful efficiency prediction

- Common criteria for successful prediction of product efficiency.
- Methodology for selecting representative input regions for accurate simulation of real-world conditions.
- Aspects of successful prediction of product efficiency by manufacturing technology factors and usage conditions.
- Building a specific type of influence function for efficiency prediction.
- System efficiency prediction from testing results of the components.
- Mathematical dependences between quantitative indices and factors that influence product efficiency.
- Mathematical description connection (during design processes) of field influencing factors with ART/ADT results of the product.

The detailed discussions of the above components can be found in [1–4].

3.3.1 Criteria of Successful Prediction of Efficiency by Results of Accelerated Reliability/Durability Testing

One can use the results of accelerated reliability/durability testing, **if it is based on accurate simulation of real-world conditions** that reflect the working environment where the equipment is expected to operate.

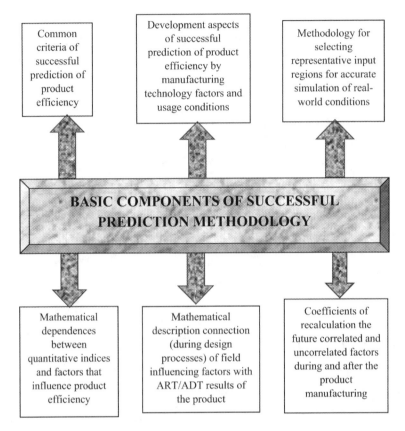

Fig. 3.9 Common scheme of methodology for product's efficiency successful prediction

But one must be sure that the prediction will be correct (if possible, with a given accuracy).

Achieving this goal can be had by using the following solution. It can be seen in more detail in references [1–3].

The problem is formulated as follows: there is the system [results of use of the current equipment in the real world] and its model [results of accelerated reliability/durability testing of the same equipment during design and manufacturing]. The quality of the system or its component can be estimated by the random value φ using the known or unknown law of distribution $F_S(x)$. The quality of the model can be estimated by the random value ϕ using the unknown law of distribution F_M. The model of the system will be satisfactory if the measure of divergence between F_S and F_M is less than a given limit Δ_g.

After testing the model, one obtains the random variables φ_1: $\varphi_1^{(1)}, \ldots \varphi_1^{(n)}$. If one knows $F_S(x)$, by means of $\varphi_1^{(1)}, \ldots, \varphi_1^{(n)}$, one needs to check the null hypothesis H_0, which means that the measure of divergence between $F_S(x)$ and $F_M(x)$ is less than Δ_g. If $F_S(x)$ is unknown, it is necessary also to conduct a test of the system. As a result of this testing, one obtains the outcomes of inputs from random variables φ: $\varphi^{(1)}, \ldots, \varphi^{(m)}$. For the above two samplings, it is necessary to check the null hypothesis H_0 that the measure of divergence between $F_S(x)$ and $F_M(x)$ is less than the given Δ_g. If the null hypothesis H_0 is rejected, the model needs updating, i.e., to look at more accurate ways of simulating the basic mechanism of real-world conditions for performing accelerated reliability/durability testing.

The estimation of the measure of divergence between $F_S(x)$ and $F_M(x)$ is done using a multifunctional distribution and depends on a competitive (alternate) hypothesis. The practical use of this criterion depends on the type and forms of this functional distribution. To obtain an exact statistical distribution that tests the correctness of the hypothesis H_0 is a complicated and unsolvable problem in the theory of probability. Therefore, here the upper limits are shown for the studied statistics, and their distributions are found, so that the level of values will be increased. Thus, any explicit discrepancies can be detected.

Let us consider the situation when $F_S(x)$ is known. For details of this concept, see Refs. [9–12]. First, we will take as the measure of divergence between the functions of distribution $F_S(x)$ and $F_M(x)$ the maximum of modulus difference:

$$\Delta\left[F_M(x), F_s(x)\right] = \max_{(x)<\infty} / F_M(x) - F_s(x)] /$$

We understand that H_0 is the hypothesis that the modulus of difference between $F_N(x)$ and $F_S(x)$ is no more than the acceptable level Δ_g, i.e.,

$$H_0 : \max_{(x)<\infty}\left[F_N(x) - F_S(x)\right] \leq D_g$$

where $F_N(x)$ is the empirical function of distribution.
Against H_0 one checks the competitive hypothesis:

$$H_1 : \max / F_N(x) - F_S(x) / > D_g$$

The statistic of the criterion can be given by the formula:

$$D_n = \max_{(x)<\infty} / F_N(x) - F_S(x) /$$

Practically it can be calculated by the following formula:

$$D_n = \max_{1 \leq m \leq n}\left\{\max\left[\frac{m}{n} - F(\eta_m)\right], \max\left[F(\eta_m) - \frac{m}{n}\right]\right\}$$

It is very complicated to find the distribution of this statistic directly [10]. The $D_n \to \Delta_g$ as $n \to \infty$.

Therefore, it is necessary to look for the distribution of random value $\sqrt{n}(D_n - \Delta_g)$.

Let us now give the upper estimation which can be useful for practical solution of our problem:

$$D_n = \max_{(x)<\infty}\left[F_n(x)-F_S(x)\right] = \max\left[F_n(x)-F_M(x)+F_M(x)-F_S(x)\right] \leq \max_{(x)<\infty}\{/F_n(x)/$$
$$F_M(x)/+/F_S-F_M(x)/\} \leq \max_{(x)<\infty}/F_n(x)-F_M)/+\max_{(x)<\infty}/F_M(x)-F_S(x)/$$

If hypothesis H_0 is true:

$$\max_{(x)<\infty}/F_M(x)-F_S(x)/\leq \Delta_g$$

Therefore

$$D_n \leq \max_{(x)<\infty}/F_n(x)-F_M(x)/+\Delta_g$$

or

$$\sqrt{n}\left(D_n-\Delta_g\right)\leq \sqrt{n}\max_{(x)<\infty}/F_n(x)-F_M(x)/ \qquad (3.1)$$

Here, if $F(x)$ is the probability of work without failure, then n is the number of failures.

Let us mark $\max/F_n(x)-F_M(x)/$ as D_n. This random value $\sqrt{n}D_n$ limited by $n\to\infty$ follows Kolmogorov's law [9]. Therefore:

$$P\left\{\sqrt{n}\left(D_n-\Delta_g\right)<x\right\}\geq k(x)$$

or

$$P\left\{\sqrt{n}\left(D_n-\Delta_g\right)\geq x\right\}<1 \quad k(x)$$

where $K(x)$ is the function of Kolmogorov's distribution.

As a result of research, the next right of use of the Kolmogorov's criterion was obtained in the following manner:

First, one calculates the number $\sqrt{n}\,(D_n-\Delta_g)=\lambda_0$.

Then:

$$P\left\{\sqrt{n}\left(D_n-\Delta_g\right)\geq \lambda_0\right\}<1-k(\lambda_0)$$

If the difference $1-k(\lambda_0)$ is small, then the probability $P\{\sqrt{n}\,(D_n-\Delta_g)\geq \lambda_0\}$ is also small.

This means that an improbable event occurred, and the divergence between $F_n(x)$ and $F_S(x)$ can be considered as a substantial, rather than a random, character of the studied values and Δ_g. Therefore, the conclusion is max/$F_S(x) - F_M(x)$/>Δ_g if (x) < ∞ is assumed.

If the level of value of this criterion is higher than $1 - k(\lambda_0)$, the hypothesis H_0 is rejected. If $1 - k(\lambda_0)$ is large, it does not exactly confirm the hypothesis, but by a small Δ_g we can practically consider that the testing results do not contradict the hypothesis.

Then, the problem was solved through Smirnov's criterion.

We ca also use Klyatis's criteria that are modifications of Kolmogorov and Smirnov's criteria.

There is the rule of this criterion to use here:

A.   One calculates the actual number $\sqrt{na}\,[R_n(a, 1) - \Delta_g] = \lambda_0$.
B.   In that case:

$$P\left\{\sqrt{na}\left[R_n\left(a,1\right) - \Delta_g\right] \geq \lambda_0\right\} < 1 - L\left(\lambda_0\right)$$

C.   If $1 - L(\lambda_0)$ is small, therefore the probability $P\{\sqrt{na}\,[R_n\,(a, 1) - \Delta_g] \geq \lambda 0$ will also be small.

This means that the difference between $F_n(x)$ and $F_S(x)$ is significant.

Then, we will take the maximum difference as a measure of divergence:

$$\Delta\left[F_M\left(x\right), F_S\left(x\right)\right] = \max_{F_S(x) \geq a} \frac{F_M\left(x\right) - F\left(x\right)}{F_S\left(x\right)}$$

This problem can also be solved by the method analogous to the previous solution. The rule of use of this criterion is as follows. One calculates the actual number with the following formula:

$$\sqrt{na}\left[R^+_n\left(a,1\right) - \Delta_g\right] = \lambda_0.$$

Then:

$$P\left\{\sqrt{na}\left[R^+_n\left(a,1\right) - \Delta_g\right] \geq \lambda_0\right\} < 2\left[1 - \Phi\left(\frac{\lambda_0 \cdot \sqrt{a}}{\sqrt{1-a}}\right)\right]$$

If $2\left[1 - \Phi\left(\dfrac{\lambda_0 \cdot \sqrt{a}}{\sqrt{1-a}}\right)\right]$ is small, it means that hypothesis H_0 is rejected and then by analogy to previous solutions.

The same applies to the competitive hypothesis H_1.

The variant was also solved when $F_S(x)$ was unknown.

In this case one provides an accelerated reliability/durability testing of the system. As a result, one realizes the random value φ of system reliability $\varphi^{(1)}, \ldots, \varphi^{(m)}$, and can use these realizations to ~build empirical functions of the distribution $F_m(x)$. By means of the empirical functions of distribution $F_n(x)$ and $F_m(x)$, one must establish whether the studied random value relates to one class or not, i.e., will the divergence between the actual functions of distribution $F_n(x)$ and $F_S(x)$, by a certain measure, be less or more than the given tolerance Δ_g.

Conclusion

1. The engineering version of the obtained solution is that the upper estimation of statistical criteria of correspondence, for some measures between the functions of distribution of studied characteristics of efficiency (and its components), was created in ART/ADT conditions and in real-world conditions. This can be useful for efficiency prediction as well as for solving other engineering problems (reliability/durability or any other efficiency component development and improvement, etc.).

The mathematical version of the obtained solution is that approximate Klyatis's criteria as modifications of Smirnov's and Kolmogorov's criteria by divergence ($\Delta_g < 0$) were obtained for comparison of two empirical functions of distribution by measurement of Smirnov's divergence:

$$\Delta\left[F_S(x), F_M(x)\right] = \max_{(x)<\infty}\left[F_M(x) - F_S(x)\right]$$

and Kolmogorov's:

$$\Delta\left[F_S(x), F_M(x)\right] = \max_{(x)<\infty}/ F_S(x) - F_M(x)$$

In Smirnov's criterion by zero hypothesis:

$$\max_{(x)<\infty}\left[F_M(x) - \overset{Ü}{F}_m(x)\right] < \Delta_g$$

By competitive hypothesis:

$$\max_{(x)<\infty}\left[F_M(x) - F_m(x)\right] > \Delta_g$$

If $\Delta_g = 0$, we have Smirnov's criterion. An analogous situation also applies for Kolmogorov's criterion.

The difference between both versions is that in the measure using Smirnov's criterion, one takes into account only the regions (the oscillograms of loadings, etc.) where $F_S(x) > F_M(x)$ and one looks for maximum differences between them.

In measuring with Kolmogorov's criterion, one considers the maximum differences for all regions by modulus. The consideration of both criteria makes sense because Smirnov's criterion is easier to calculate, but does not give the full picture of divergences between $F_S(x)$ and $F_M(x)$.

Kolmogorov's criterion gives a fuller picture of the above divergence, but is more complicated to calculate.

Therefore, the choice of the better criterion for a specific situation must be decided according to the specific conditions of the problem to be solved.

Let us show the solution obtained with a practical example. From the field a sample of 102 failures ($m = 102$) in car trailer transmissions were recorded. After accelerated reliability/durability testing (ART/ADT), 95 failures were obtained [($n = 95$), Δ_g is 0.02].

One builds the empirical function of distribution of the time to failures in the field $F_m(x)$ *by* the intervals between failures, and one builds by intervals between failures during ART/ADT of the empirical function of distribution time to failures $F_M(x)$. As we can see, this is the last variant to be considered.

If we align the graph $F_M(x)$ (Fig. 3.11) and the graph $F_m(x)$, we will find the maximum difference between $F_M(x)$ and $F_m(x)$.

For achieving this goal, we can draw the graph $F_m(x)$ on the transparent paper, and it is simple to find the maximum difference $D^+_{m,n} = 0.1$. In correspondence with [11], we have $\lambda_0 = 0.98$:

$$\text{The } k = \frac{m}{N} \approx 1, \text{therefore } F_x :$$

$$F_x(x) = 1 - e^{-2x^2}\left[1 + x\sqrt{2\pi}\cdot\Phi(x)\right]$$

After substituting $\lambda_0 = 0.98$, we obtain $F_x(0.98) = 0.6$. Therefore, $1 - F_x(0.98) \approx 0.4$. So, $1 - F_x(0.98)$ is not small and the hypothesis H_0 can be accepted. Therefore, the divergence between the actual functions of distribution of time to failures of the above transmission details for the tested car trailer in field conditions and in ART/ADT conditions by Smirnov's measure is within the given limit $\Delta_g = 0.02$ (Fig. 3.10). This gives the possibility for successful prediction of reliability of the car trailer transmissions using the results of this testing.

3.3.2 Development of Techniques for Product Efficiency Prediction on the Basis of Accelerated Reliability/ Durability Testing Results

The problem to solve here is the successful prediction of product efficiency if one takes into account the **full range of input** influences, **as well as human factors and safety problems** from the complex input factors on efficiency.

The typical situation in practice for engineering during a design process is using a small sample size (from five to ten specimens of each component) with only two to five possible failures included.

Usually, it is assumed that the failures of the system components (as well as efficiency components) are statistically independent.

Fig. 3.10 Evaluation of the correspondence between functions of distribution of the time to failure of a car trailer's transmission detail in the field and in the ART/ADT conditions

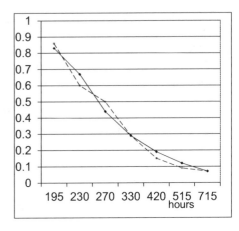

The proposed approach is very flexible and useful for many different types of the product such as electronic, electromechanical, mechanical, and others.

One can see the full solution of this problem in Sects. 3.1 and 3.2.

3.3.3   Basic Concepts of Efficiency Prediction

The best methodology of efficiency prediction cannot be useful for practical engineers if it is not connected with effective techniques and equipment that get accurate initial information for this prediction.

The techniques and equipment of accelerated reliability/durability testing can give this information if step-by-step testing, as shown in Sect. 4.3, is performed:

1. Building an accurate model of real-time performance. This model will be used for testing the product and, as a result, to study the degradation that occurs in anticipated field use.
2. Test the mechanism over time and compare it with the real-life degradation of this product. If these degradation mechanisms differ by more than a fixed limit, as described in the formulas above, one must improve the model's real-time performance.
3. Make real-time performance forecasts for efficiency prediction using these testing results as initial information.

Each step can be performed in different ways, but efficiency can be predicted accurately if researchers and engineers use the above concept correctly.

Step 1 can be executed if one understands that in real life the efficiency of the product depends on a combination of different interacting input influences, such as shown in [2], as well as safety and human factors. In this case the simulation of real-life input influences will be as complicated as in real life. For example, for a mobile product, one needs to use an interactive multi-axis vibration model in combination with multi-environmental and other groups of testing.

To solve step 2, one must understand the degradation processes of the product and parameters of these processes. The results of the product degradation process include data on the electrical, mechanical, chemical, thermal, radiation, and other variable effects. For example, the parameters of the mechanical degradation process are deformation, crack, wear, creep, etc. In the real world, different processes of degradation act simultaneously and in combination.

Therefore, accelerated reliability/durability testing includes simultaneous combination of different types of testing (environmental, electrical, vibration, etc.) [17], with the assumption that the failures are statistically independent. The degradation process of the product in real life must be similar to this process during ART/ADT.

To solve step 3, the efficiency prediction technique must be developed. For solving this problem, the factors that must be considered involve specific manufacturing and field conditions.

3.3.4   Other Methodological Aspects of Successful Prediction

One can read in detail other methodological aspects of successful efficiency and its component prediction in author's books [2, 3, 5].

The essence of these aspects is the following:

- **Successful Prediction of Product Efficiency by Considering Coefficients of Recalculation That Depend on Manufacturing Technology Factors and Usage Conditions.**

The successful prediction methodology helps to predict components of efficiency during:

- Early design
- The following design steps
- The manufacturing process
- Usage

It helps to compare the planned and factual indices of the mentioned factors to the requirements and define the requirements more precisely. It also helps to provide the necessary work for accelerated product development and improvement through all steps of design and manufacturing.

Usually, quantity indices of efficiency components are obtained from experimental testing of product specimens which do not take into account the specific of manufacturing processes.

Therefore, for successful prediction of produced product efficiency with taking into account manufacturing technology with changing this during the time, testing should be done before manufacturing, through specific mathematical modeling.

Then, the results can be corrected after providing manufacturing process. This problem solution is described in the book [2].

- **Building a Specific Type of Influence Function for Efficiency Prediction**
  The following building a specific type of influence factors can help to evaluate:

- The level of combined influence of the uncorrelated generalized factors of manufacturing technology and other real-world conditions (which cannot be take into account during design process)
- The level of separate groups of the above factors on product efficiency
- The level of influence the most important manufacturing factors.

This solution process with an example is also demonstrated in author's book [2].

- **Basic Methodological Aspects of Quality Prediction**
  In a broad sense, quality directly includes reliability, durability, maintainability, and many other factors. But practically speaking, especially in the United States, quality usually includes quality assurance, quality control, quality auditing, and quality inspection. Usually, the specific of quality consider these components only for a short time during design or manufacturing with corresponding measurements.
  The basic elements of considered quality are:

- Quality control
- Quality assurance
- Inspections
- Reliability/durability/maintainability
- Quality engineering
- Quality audit
- Procurement quality
- Metrology measurement
- Administration

For successful prediction of basic elements of quality, one needs to provide complex qualitative analysis, including quality assurance and quality control of factors that influence product efficiency during design, manufacturing, and usage phases of service life.

One can see this solution with examples in the author's book *Accelerated Quality and Reliability Solutions* [2].

- **Methodology for Selecting Representative Input Regions for Accurate Simulation of Real-World Conditions**
  One can see this methodology in Chap. 4. Prior to the author's scientific studies, there were no methodologies for selecting a real-world representative region that could be used for simulation before ART/ADT providing. As he discovered, the basic problem is follows. In real-world conditions, most products are used in a wide variety of environmental zones as found in different geographic land masses. Similar differences of real-world conditions are found in air, space, and water environments.

As a result, the efficiency components in the different regions will be vastly different. Therefore, we must establish representative regions of input processes that correspond closely to the different field conditions.

Then, simulate this representative region for providing an accurate ART/ADT technology. If there is significant difference between some regions, we need to decide how it is necessary to simulate both of them.

- **System Efficiency Prediction from Testing Results of the Components**

This problem can be solved when testing a complete system that has either a high cost or insufficient time to testing, especially at the beginning of product development. To achieve this goal, a multivariate Weibull model can be used for the prediction of efficiency (or its components) from testing results of the product components. It is assumed that the system (complete machine) consists of N components (units and details). For each of them, the failure is statistically independent, and the Weibull lifetime distribution is used with scale parameter β and shape parameter α_{ij}, $I = 1, \ldots N$. For each component, the testing results are obtained using sensors to provide the data.

The author with his colleagues has offered an algorithm for calculating the lower confidence bounds with a given confidence level of system efficiency (and its components) prediction [14]. The initial information for this practical prediction for any given time can be obtained as a result of ART/ADT.

- **Durability Prediction with Consideration of Expenses and Losses**

If one uses ART/ADT in accordance with Chap. 4, he/she can obtain accurate initial information for durability prediction. During the testing period, the degradation of the test subject and the expenses used for this subject during the time are approximately identical, considering the correlation (taking into account the accelerated coefficient) of ART/ADT results and the real-world results. The aforementioned is true, because ART/ADT needs accurate simulation of the real-world conditions. One group of expenses is evaluated by ART/ADT and another group by periodical field testing, which includes the comparison of test subjects who have executed different volumes of work (hours, miles, etc.) during ART/ADT.

The total expenses occurred during the use of test subjects consist of three basic components:

- Expense related directly to the work of the test subject
- Expenses that depend on loss of the product during normal work of the test subject
- Expenses that depend on stopping the test subject work if there are failures or increased degradation of test subject (increased degradation and stopping of work means there is increased product loss and associated costs are incorrected)

One can see this problem solution with practical example in Ref. [2].

- **Prediction of the Product's Spare Parts**

If one has the number of basic components and time of its serviceability distributed by unspecified law, it is possible to find the correct number of spare parts for a given period of usage. Thus, the ability of the support system to provide spare parts at given interval of time will not be less than given probability. The probability of the problem spare parts is obtained by how we define reliability (or other efficiency component) of success, or the probability that the system will perform its intended function at a certain given time. In other words, the number of elements of failures will be no more than m, whereas the number of serviceable elements (details) is greater than or equal to n.

For any given probability, the problem of finding the number of spare parts was solved. This solution can be seen in Ref. [15].

- **Successful Prediction of Some Financial Components (Life Cycle Cost, Profit, and Recalls)**

Financial prediction is an important part of a system's successful prediction. The solution of financial problems can be found and described in Refs. [2, 4].

It was stated that financial efficiency components, such as life cycle cost, profit, recalls, and others, depend on the accuracy of the interconnected efficiency components (reliability, durability, maintainability, safety, etc.) prediction. Recalls are related to financial parameters, because the recall cost directly depends on the above components, reputation in the market, and sales volume. The above accuracy depends on how accurate were the simulations, how accurately one provides ART/ADT, and, most importantly, how accurately did the initial information provided for prediction reflect the real world's usage environment.

References [2, 4] show how these efficiency components can be practically predicted.

Let us consider briefly the author's approach to prediction of life cycle cost (LCC).

First, reliable initial information must be had for prediction, which is based on accurate real-world simulation and accurate ART/ADT technology, as described in this book.

The product LCC can be described in terms of costs incurred ranging from the cost of inception to that incurred during its service life. This life cycle has three stages:

1. Research and process planning
2. Design and manufacturing
3. Entire usage and ownership, including operation, maintenance, repair, and other during the usage

The first stage is research, planning, and process of pre-design development. Starting this stage will be some form of market analysis of a commercial product that ensures there will be a demand for the product. For military or other nonprofit organizations, a review is made to determine the need for the system that cannot be fulfilled by any other means. The importance of these reviews cannot be overstated as a decision to go forward with the development of the system will encumber years of cost investments. For example, KIA Motors markets their cars with a 10-year maintenance guarantee. Clearly, they believe the market will be receptive to this maintenance-free appeal. Consequently, their cars must be well designed for an elevated level of reliability that will require a low cost of maintenance. Once a commitment to proceed with a system's development, the decisions made by researchers and technical managers during this stage directly affect every efficiency component (reliability, durability, and others). It is here where prediction is critically important. Every design failure that can be found and corrected through such prediction will be result in significant life cycle cost savings.

The second stage is design and manufacturing that is the actual making of the product. This includes the design process directly, material selection, scheduling and routing procedures, testing, quality audits, development and using manufacturing technology, distribution to the customers, and market outlets. To the manufacturer, the dominant dimension here is the conformance to quality, which is assessed and tracked through quality control procedures and inspections.

The third stage of LCC is the product usage and ownership, which is traditionally considered to be the product life cycle, and starts when an item is received by a customer. These costs will continue through the retirement of the system. These costs comprise the vast preponderance of the LCC, and hence, the work done in the first two stages will result in a consequent savings in this third stage.

The total product life is the time spent in developing and producing the product plus the usage stage. The usage study covers the entire span of usage and ownership and includes operation (including cost of product, insurance, fuel, etc.), maintenance, any resale, complaints, recalls, and others. Usually, the cost of this stage is more than 60–70% of LCC.

For example, as was written in *KIMM (Korea). Expenses During Life Cycle Cost of Machinery. Introduction to Reliability Assessment Center*, expenses (weight) incurred during the life cycle of machinery:

- Design – 1
- Engineering – 10
- Manufacture – 100
- Usage – 1000

- **About Recall Prediction Methodology**
  If one known the following:

1. Methodology to successfully predict reliability, durability, quality, maintainability, LCC, and other efficiency components (see earlier discussions in this chapter [1, 2]
2. Strategy, methods, and equipment for the specific product ART/ADT (see Chap. 4), he/she can predict changes during service life interconnected with safety, as written in this book and in more detail in [2], and the author's papers submitted at the SAE World Congresses.

As a result, one could for each specific product:

- Provide successful prevention and prediction of recalls, resulting in improved reliability, durability, and safety, including the influencing factors during design, manufacturing, and usage

- Finally, dramatically decrease recalls and other returns through the elimination of their causes during ART/ADT (studying quality, reliability, durability, and maintainability problems, and eliminating most of them)

3. The key factor of this approach is the accurate simulation of the real-world conditions for ART/ADT, as described in this book, and in more detail in [1, 2].

Companies in automotive, as well as in some other industries, have not yet applied this strategy for entire machines, as well as their components (units and details), on a wide enough and in-depth basis. (Until now most reasons are described in this book.) The key reason is that companies do not fully appreciate the large financial opportunity gained from applying this approach.

- **About Profit Prediction Methodology**

One can successfully predict profit if the performance components can be successfully calculated using ART/ADT results, as described in Chap. 4 (and in [1, 2]), and using the methodology of prediction of efficiency components. The LCC, recalls, stop-ships, retrofits, other returns, and expenses can be easily calculated using the methodology mentioned earlier and the ART/ADT results.

One can easily calculate also other components of profit by using the standard methods with a company's real expenses. Finally, using this book approaches, anyone can predict the profit for each product (technology) given knowledge of all of the expenses, losses, and earnings from the sale of the product.

Conclusion

Using the aforementioned prediction methodology and the strategy leads to a dramatic improvement in the financial costs in the design, production, distribution, and follow-on support of a product. During ART/ADT, one can provide events for reliable prediction of interacting reliability, durability, safety, quality, maintainability, and other components of product efficiency. To achieve this goal, one needs to analyze the causes of product degradation (failures) and eliminate them if possible. This helps to successfully predict and eliminate many problems related to the service life of the product during research, design, and manufacturing.

Finally, one can successfully predict the profit, as well as other economic indices of the efficiency as described in this book and in more detail in Refs. [2–4].

3.4 Improving Engineering Culture for Efficiency Successful Prediction

Cultural engineering is the conceptual approach to product development planning and management that accounts for changing strategies that deal with issues raised by culture and social development in diverse contexts.

In other words, cultural engineering is about systems, processes, and alternatives, and the formulation of creative solution to challenges in the development of cultural institutions and the promotion of people's participation in cultural life.

Problems in economic viability, sustainability, technology, and the overall sustainability advances and the overall aesthetic values of our modern culture now demand a comprehensive set of skills that combines and manages many diverse fields that make up a well-designed project. Call it synergy, mechatronics, cognitive or systems engineering, it must something that works.

Commonly, the improvement of culture inside and among partnered companies and organizations means the establishment of positive relationships with everyone that result in improved communication, reduced distress, and in general a healthy work environment. Of done right, this will result in committed work team working efficiently toward customer satisfaction through the provision of high-quality goods and services.

The idea of engineering culture is important, because many engineers orient their identities and careers to their occupation rather than to their organizational communities [21]. For engineers in particular, paying attention to an engineering culture may be more telling about engineering practices and values than examining the companies for which the engineer works. This study has the potential to advance our understanding of how and why engineers interact with one another the way they do.

Engineering managers should be taught how to select team members, assign team roles, build cohesive unity, and assess progress toward identified goals. Such an analysis relies on the structure of teamwork itself to help engineers work together in teams. Current studies agree that teamwork structures are important for helping individuals to interact productively. Specifically, understanding cultural mores, values, and practice enables us to see how interactivity among engineers really works.

Scholars, practitioners, and employers alike are noticing that engineers are not well trained to work with other individuals to bring a project to fruition. This fact is critical, because the scope of most projects is such that they require multiple engineers to work on them simultaneously a skill most engineers lack.

Across a wide variety of literature, researchers consistently identify similar values and practices that characterize an "engineering culture." Engineers themselves are aware that they belong to a professional culture that sets explicit guidelines for what it means to be an engineer. The amazing coherence and persistence of this engineering culture suggests that there is a certain mythos surrounding it.

Gideon Kunda [22] offers a critical analysis of an American company's well-known and widely copied "corporate culture." The company's management, Kunda reveals, uses a variety of methods to promulgate what it claims is a nonauthoritarian, informal, and flexible work environment that enhances and rewards individual commitment, initiative, and creativity while promoting personal growth. In a conclusion written especially for his latest studies, Kunda reviewed the company's fortunes in the years following the publication of his earlier investigation. During this review, he revaluated his early arguments and explored the relevance of corporate culture and its management today. His latest studies demonstrate clearly that these pervasive efforts mask an elaborate and subtle form of normative control in which the members' minds and hearts become the target of corporate influence. Kunda carefully dissected the impact this form of control has on employees' work behavior and on their sense of self. His findings were not good, but rather the management practices he saw were determined to productive work.

A systems of systems approach or systems engineering relates well to an engineering culture and is simply stated, as a system which is an integration of three elements composed of:

- People
- Products
- Processes

that provides a capability to satisfy a stated need or objective.

As we can see from the above, an engineering culture consists of not only the relationship among the people but also the consideration the products and processes being developed. Not understanding this leads to narrow thinking of all levels of engineering, beginning with the president (CEO) of an industrial company and continuing down through all levels management to the engineers and other specialists of the company. All these people should have one final goal, making their company successful in the market, and if the culture is poor, it won't be achieved.

One of the basic problems of upper management is strategic thinking, but this means more than simply process efficiency. Effectiveness must be part of equation. To achieve both goals, a people-oriented corporate engineering culture that is based on trust and integrity must be established and maintained through every success and calamity. It must be honestly and consistently applied if the lower levels of professionals are to practice it. Let upper management be seen as hypocrites here and the entire corporate culture will be one of cynicism. However, if upper management is seen as honest, then the lower-level staff will accept the corporation's engineering culture.

The same concept relates to various models of a product, which are currently produced by a company.

They are interconnected, because the actions of each are like the links of a chain. These interconnections relate to people, products, and processes of companies-producer full vehicles, as well as their suppliers. In fact, one can see that, beginning from top management, people think mostly about their narrow responsibility and less about how their responsibilities are interconnected in the real world with other areas of company activity. Correcting this misperception should be a key responsibility of top management.

For example, often a technical director who is responsible for human resources does not think about how his area relates to other areas of this company. Therefore, he cannot solve the basic human problems of his or her division, because they are considered separately from the problems of other interconnected areas of the company.

The above is similar to testing, where, for example, a person responsible for corrosion testing may think only about this type of testing in relation to chemical pollution. The engineer may not consider that corrosion of the product also depends on vibration, deformation, dust pollution, etc. Moreover, engineers also do not realize that degradation and, finally, failure of the product, in addition to corrosion, depend on wear, cracking, solar radiation, human factors, and many other factors. In this case, improving engineering culture means understanding and considering all of the above factors.

One more problem in improving engineering culture relates to testing costs. Usually, when one compares different testing approaches, the cost of each approach is directly compared, and the cheaper testing approach is selected.

In this case, one directly compares only the cost of the testing approaches. This shortsightedness demonstrates a poor engineering culture, because the selection's influences on the life cycle cost of the selecting testing have not considered. Sometimes, the selection does not influence subsequent processes' costs, but, often, will portray an inaccurate simulation of the real world, thereby influencing subsequent processes' costs that lead to unpredictable degradation, failures, and, finally, recalls and loss of profit. By contrast, improved engineering culture leads to using life cycle cost for comparing different testing approaches. In practice, using accelerated reliability/durability testing (ART/ADT) technology leads to lower life cycle cost, especially with reducing recalls [23].

Therefore, development of engineering culture must go through a comprehensive development of many interacting factors. Results like the above come from a poor prediction of product efficiency that is based on poor initial information, thereby leading to increasing recalls and decreasing profit.

Successful prediction of reliability, durability, safety, quality, life cycle cost, profit, recalls, and many other components of efficiency cannot be done accurately without development of an effective engineering culture that connects the interacting components for a comprehensive, integrated body of test results.

Sufficient efficiency prediction of the system requires knowledge of all the components in a chain of interacting components as seen in Fig. 3.11.

Failure to adhere to the above concept is the basic reason why automotive and other industries cannot prevent recalls.

Prediction is inaccurate, when no comprehensive methodology for the successful prediction of product efficiency is selected. Often product's efficiency is obtained because of separate vibration, or corrosion, or other types of testing. As a result, the final answers are disjointed and not integrated.

As was mentioned, the author developed successful prediction methodology, which consists of the following basic components:

- Establishing the criteria that indicate a successful prediction of performance components after testing
- Writing a clear mathematical description of dependencies between quantitative indices and factors that influence prediction
- Writing a clear mathematical description of dependences of the connecting influence factors on the testing results for the product specimen during the design process
- Developing the coefficients for recalculating the future correlated and uncorrelated factors and quantitative indices of performance components during and after manufacturing

Fig. 3.11 Scheme of successful component prediction, which is similar to a chain of interacted links

This methodology also includes description of qualitative and quantitative level of performance.

Prediction is also inaccurate when it is based on:

- A methodology not connected closely with the sources of accurate initial information for this prediction calculation;
- Using traditional approaches of accelerated life testing (ALT) data where the degradation (failure) process differs substantially from the product's degradation process during service life under real-world conditions. It is not based on the technology of accelerated reliability/durability testing (ART/ADT), as a source of initial information and as a key factor for successful prediction.

Why is this so?

There are two basic groups of causes why ART/ADT is not often used for this situation as indicated in Fig. 3.12:

- Cultural group of causes:

    - Many industry professionals do not want to be held responsible for poor analysis and why their company's profit much less than what was planned during the design and manufacturing
    - Poorly defined role of the testing in the development and manufacturing processes
    - Why the system of comparison testing conditions with real-world conditions is so poor

- The basic technical causes of the above:

    - Not enough development of the theory about accurately simulating real-world conditions
    - Not enough development of the strategy for accurately simulating field conditions

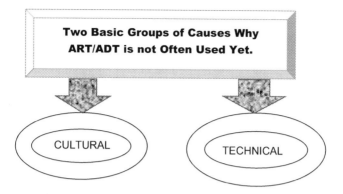

Fig. 3.12 Two basic groups of causes why ART/ADT is not often used yet

- Undeveloped methodologies for accurate simulation of field conditions in the laboratory (or proving grounds) that include full input influences, safety, and human factors that affect the efficiency of the products
- Undeveloped equipment for accurate simulation of the real-world conditions in the laboratory
- The knowledge of advance testing, including reliability and durability testing, obtained from the literature, is often poor.

The above are the basic reasons why recalls have not decreased from year to year.

Improvements in the engineering culture in testing have been lacking in the following ways:

First is use of laboratory stress testing (often called accelerated life testing – ALT). This approach began many years ago. For this testing, one simulates single input influences or several input influences (temperature, humidity, pollution, or others). This contradicts conditions in the real world, where many input influences are acting simultaneously and in combination with each other.

Next was the development of improved accuracy of the simulation of these separate components of field parameters, but they still did not mirror real-world conditions and hence the test results were flawed.

In the 1950s engineers understood that separate simulation of one field input influence (parameter) did not reflect field conditions where many parameters act in combination. From that time engineers began to study and use combined accelerated stress testing such as temperature, humidity, and vibration in special test chambers.

This testing process rapid development began with electronic products.

Then, in the 1990s, advanced companies, first, for electronic product testing, developed test chambers with a simultaneous combination of multiple parameters. These test chambers usually combined the simulation of temperature, humidity, vibration, and input voltage.

From this time the improvement of the engineering culture in the development of accelerated life testing (ALT) moved in two directions.

One direction moved to highly accelerated testing methodologies: HALT (highly accelerated life testing), HASS (highly accelerated stress screening), and AA (accelerated aging). The basis of these types of accelerated testing is the simulation of two factors (commonly temperature and vibration), with the level of each of these factors being much higher than in the field. For example, in the automotive industry, where the maximum field temperature is 70 °C, the test chamber temperature may be as high as 120 °C. Likewise the minimum field temperature is −40 °C, but the test chamber temperature may be as low as −100 °C.

This is a blind method (approach) of testing because the physics of degradation process has been altered. Therefore, one inaccurately simulates the real-world conditions. Changing the physics of degradation process in the laboratory, in comparison with the field, changes the corresponding time to failure, which influence the accelerated coefficients for different details of the vehicle. As a result, one cannot accurately calculate the accelerated coefficient until after experimental results of the service life for each separate subject have been obtained.

A similar situation is seen with vibration [23].

And second, one cannot evaluate accurately the accelerated coefficient for the entire vehicle or unit that consists of many details with different accelerated coefficients.

The fundamental drawback of this approach is its inability to successfully predict reliability, durability, maintainability, quality, and other efficiency components during the product's warranty period or service life. Somebody calls this "modernized simulation" of field inputs.

Therefore, the long-term effectiveness of this type of testing is not better than the traditional (earlier) type of testing.

Also, the level of development of HALT from the ART/ADT level is not closer than that of traditional separate testing methodologies influenced by ART/ADT. HALT and HASS are popular by users, because they are simple and inexpensive. But such professionals often do not consider the cost of subsequent processes; therefore, it is a blind path for the solution of successful prediction of the product's quality, reliability, durability, maintainability, recalls, profit, life cycle cost, and other components of efficiency during the extended period of a product's service life.

Often people incorrectly call combined types of testing or vibration, or proving ground testing, as durability testing. This is not right, because durability testing, like reliability testing, needs accurate simulation of real-world conditions, i.e., simulation of the full input influences of the real-world (multi-environmental, mechanical, electrical, and others) plus human factors and safety. More developments have entered the accelerated reliability/durability testing, which is described in Chap. 4 of this book and in more detail in the Refs. [23–29].

There is one more problem that relates to the improvement of engineering culture. During the past 20–30 and more years, there has been an accelerated effort to move product manufacturing to the lowest cost locations. In the case of the complexity of products and supply chains, this often leads to suboptimization of the supply chain system. Suboptimization may occur when the pursuit of the lowest segment price adds cost to the system's (whole vehicle) life cycle.

A systems engineering approach to analyzing an entire supply chain for nonvalue adding activity – some of which is introduced through the pursuit of low-cost suppliers – can help to optimize a supply chain and result of cost saving and improvement in component performance.

To understand a supply chain, it is necessary to evaluate many characteristics of its supplier. Eight characteristics of products and services in a lean environment – known as the eight rights – must be evaluated and understood. These eight rights include [30]:

- The right product
- The right quantity
- The right conditions
- At the right place
- At the right time
- From the right source
- At the right price
- With the right service provided

While there is not a one-to-one relationship between the eight rights and the seven-supply chain wastes, the supply chain wastes collectively can be seen as root cause of poor performance on the eight rights.

The most common tool used to address the eight rights is a plan for every part (PEEP). PEEP is used in the planning for all new parts and suppliers. It is a holistic tool in which all supply chain performance characteristics of a purchased component are documented. A PEEP allows an organization to drill down into details of the supply chain and determine optimal methods to manage suppliers so complexity can be driven out.

The above eight rights are a subject of a PEEP in that they allow for the measurement of critical performance components of purchased parts for every shipment received. Each right is measured by the percentages of successful executions. A "perfect execution score" is delivered by multiplying together percentage of successful executions for each of the eight rights. More detail regarding this approach can be found in Ref. [31].

Improvement in engineering cultures is also discussed in the article "The Future of Quality. In fifteen years will you recognize your organization?" [31]. The authors consider quality in a broad sense (including reliability, durability, maintainability, and others). Here are salient quotes from their article [31]:

- "One future looking trend in today's leading organizations is the renewal interest in customer focus."
- "Buzzwords are emerging already for the new quality world, as predicted by quality researchers and futurists."
- "The world is changing faster than the management realizes."

- "In 2020, we'll need bifocal leadership: clear, short-range thinking and sharp action to steer through the downturns, as well as accurate vision and steady nerves to see well into the future."
- "Known & go:
  - Quality organization in 2020 will be dominated by the acronym future, which stands for fast, urban, universal, revolutionary and ethics;
  - Quality professionals will continue to be motivated by factors other than money; the evaluation of information technology and the formulation of virtual companies and demand new approaches to managing quality;
  - Forget Six Sigma. Tomorrow's focus will be on error-free performance. Quality efforts in the United States will focus less manufacturing, because that industry will represent just a fraction of our gross national product."

An important interview with quality legend Joseph M. Juran [32] relates to the improvement of engineering culture. This interview discusses the impact of manufacturing in the United States. Although given in 2005, this interview is still current. Below are some important components of Juran's answers [32]:

- "Some people think that high quality costs more. That confusion exists in many different companies. The word 'quality' has two very different meanings. One meaning is the features of the product that enables it to be sold. There, higher quality generally costs more. It takes more product research, more product development and so on. People don't even call it a cost, they call it an investment, which will bring back higher returns. That's quality on the marketing side or the income side.
- Quality on the cost side is quite different. The cost of failure includes the internal failure – scrap, rework, slow deliveries, failure to deliver on time – and the external failures – field failures, law suits, safety problems.
- A lot of CEOs believe that they are too busy to lead the quality charge, and so they delegate it. That hasn't worked very well. Leadership by the top people is an essential ingredient in getting out of that steep slope to failure.
- A lot of companies believe that getting certified to ISO 9001 solves their quality problems. That simply is not true.
- The different members from companies of different standardization bodies are not going to agree to standards that their companies are not able to meet. They are starting to change the standards, but that's at a glacial pace. It takes a long time to change international standards." Today's world is fraught with risk. A failure mode and effect analysis (FMEA) is an analytical methodology used to ensure that potential problems have been considered and addressed throughout the product and process development cycle."

An offshoot of Military Procedure MIL-P-1020, entitled Procedures for Performing, Effects and Critically Analysis, issued in 1949 [34]. Even through it is an old publication, it is still has valid concepts.

Failure mode and effect analysis (FMEA) was first used as reliability evaluation technique to determine the effect of system and equipment failures. Failures were classified according to "their impact of mission success and personnel/equipment safety" [34]. FMEA was further developed and applied by NASA in the 1960s to improve and verify the reliability of the space-program hardware. Today, the procedures called out in MIL-STD-1629A are the most widely accepted methods used throughout the military and commercial industry.

FMEA is a prevention-based, risk management tool that focuses the user or team on systematically:

- Identifying and anticipating potential failures
- Identifying potential causes for the failures
- Prioritizing failures
- Taking action to reduce, mitigate, or eliminate failures.

The real value of the FMEA is reflected in its use as a long-term, living document. It is essential that the document is owned and updated as changes are made to field design or process.

There are two types of FMEAs [33]:

- "DFMEA (Design Failure Mode Effect Analysis) – an analysis process used to identify and evaluate the relative risk associated with a particular hardware design.
- PFMEA (Process Failure Mode Effect Analysis) – a similar analysis, which relates to process design."

For improving engineering culture, one needs to learn the following top FMEA mistakes and do not use them [35]:

- Some design FMEAs drive mostly ineffective testing, while some process FMEAs drive mostly controls.
- Some FMEA teams do not have the right experts on their core teams. These team members just sit in their chairs if they show up at all meeting while not contributing to team synergy. The mistake is that FMEA teams are inadequately composed and poor participation.
- Do not provide links between FMEAs and field data or manufacturing data. A lack of integration can cause serious problems to be repeated. This mistake results is in a serious disconnection between the FMEA team and information from the field or plant.
- At least 50% of field problems can occur at interfaces between parts and subsystems or between the system and environment. Similarly, many manufacturing or assembly problems occur at the interface between operations, such as while transporting materials, receiving incoming, or shipping goods. Some practitioners miss these interfaces.
- Some organizations miss the opportunity to improve their design verification plan (PCP) bases on the failure modes or causes from the FMEA. The results are inadequate product testing or PCP and failure of the FMEA to improve test and control plans.
- Some organizations do not encourage or control proper FMEA methods. Or, they copy old FMEAs and don't adequately address changes, such as new technology.

- Defining risk while using inappropriate criteria. Failure to address all high-risk failure modes can result in potentially catastrophic problems or lower customer satisfaction.
- Some organizations mandate FMEAs, but that doesn't ensure the time spent on them is productive.

One more important aspect of engineering culture is correctly understanding the meaning of terms and definitions. For example, one could read in Carlson's article "Which FMEA Mistakes Are You Making?" [33] how an inaccurate understanding of the term "durability testing" and its definition lead to poor testing, evaluation, and prediction of the product performance.

A similar situation is with the term "real world" as described throughout this book.

As will be discussed in Chap. 5, some negative trends in development prediction and testing in engineering relate to narrow thinking during development autonomous vehicles, where one pays more attention to driving testing and ignores reliability/durability testing. This is a result of poorly developed engineering culture.

3.5   Organizational Culture as a Component of Improving Engineering Culture

Often organization development practitioners must deliver successful projects and nudge an organization forward in an improvement journey. Finding the right path involves dealing with uncertainty, because of an organization's unique culture and experience. When things are uncertain, one does not get clear answers. As a result, initial efforts may not work or the team could easily misstep. In such situations, based on research done by the author's team, our response to inevitable setbacks is the key to guiding the improvement during the journey forward.

The first setback that must be overcome is often defined as "political." Change can be seen as a personal threat that results in direct confirmation tactics to overwhelm opposition. Such advocates offer hard-selling solutions or uncompromising positions that leads to stonewalling, disengagement, and painful memories of the improvement initiative. In other words, all pain, no gain.

What does one do in this situation? Before selecting a project, gauge the organization's pulse.

Interact with all levels of management and staff, get a sense of what they do, and identify key challenges that exist.

Identify initiatives launched in the past, the reason for undertaking those initiatives, how they were managed and supported, and how different parts of the organization adapted to the changes [35].

Professionals often recommend this approach for gauging an organization's culture before fully committing to a project. Has the organization undertaking similar projects in the past? How was the experience and outcome? Developing these insights before starting work on a lean project helped us craft a better strategy by addressing three key issues:

1. Address the past bad experience.
2. Use early engagement, inquiry, problem solving, and negotiations to win over employees.
3. Frame it right.

For more detail on an organization culture, the reader is referred to Manof and Varday's article *Discovering the Right Path* [35].

Conclusions

1. The section considered why improving engineering culture is so important for engineers and managers and how it relates to systems engineering. The idea of engineering culture is important because many engineers orient their identities and careers to their occupation rather than to their organizational communities.
2. The engineering culture consists of not only the relationship of the people. It relates also to the products and processes. Not understanding this leads to narrow thinking of all levels of engineering.
3. For improving engineering culture in automotive and other industries, one has to:

   - Consider development of engineering culture for successful prediction of reliability, durability, maintainability, safety, life cycle cost, profit, recalls, and many other components of efficiency. Prediction of the system requires knowledge of all the components in the chain of interacting.
   - Understand the supply chain and to evaluate many characteristics of supplier, including eight characteristics of products and services in a lean environment (described above in this section).

4. An important interview with quality legend, Joseph M. Juran, discusses the impact of manufacturing in the United States. He explains why high quality costs less.
5. In Sect. 3.3 the author demonstrated the basic components of methodology (3.3) as the first basic aspect for successful prediction of product efficiency as they relate to reliability, durability, life cycle cost, profit, recalls, and others. Successful prediction of the product efficiency is one from many basic aspects for improving an engineering culture.

References

1. Lev Klyatis (2012) Accelerated Reliability and Durability Testing Technology. John Wiley and Sons: New Jersey.
2. Lev Klyatis and Eugene Klyatis (2006) Accelerated Quality and Reliability Solutions. Elsevier Science: Oxford, UK.
3. Lev M. Klyatis, Edward L. Anderson (2018) Reliability Prediction and Testing Textbook. Wiley.

4. Lev Klyatis (2016) Successful Prediction of Product Performance Quality, Reliability, Durability, Safety, Maintainability, Life Cycle Cost, Profit, and Other Components. SAE International.
5. Lev M. Klyatis (2020) Trends in Development Accelerated Testing for Automotive and Aerospace Engineering. Academic Press (ELSEVIER).
6. Lev M. Klyatis, Eugene L. Klyatis (2002) Successful Accelerated Testing. Mir Collection.
7. 凯耶斯 (Klyatis, Lev M.) (俄)Lev M. Klyatis著 ; 宋太亮, 方颖, 丁利平等译 宋太亮, 方颖, 丁利平, ; Lev M. Klyatis.; Tailiang Song; Ying Fang; Liping Ding. 加速可靠性和耐久性试验技术 = Accelerated reliability and durability testing technology / Jia su ke kao xing he nai jiu xing shi yan ji shu = Accelerated reliability and durability testing technology. 国防工业出版社, Beijing : Guo fang gong ye chu ban she.
8. L.M. Klyatis (1985) Accelerated Evaluation of Farm Machinery, AGROPROMISDAT, Moscow.
9. A.N. Kolmogorov (1941) Interpolation and extrapolation of stationary random sequences. Paper of USSR Academy of Science, Moscow, No. 5.
10. B.L. Van der Varden (1956) Mathematical statistics with engineering annexes.
11. E.S. Ventcel (1966) Theory of Probability, Moscow, Highest School.
12. I.V. Smirnov (1944) Approximation of distribution laws of random variables by empirical data. Papers of USSR Academy of Science, 1944, 10, pp. 179–206.
13. Lev Klyatis (2009) "Accelerated Reliability Testing as a Key Factor for Accelerated Development of Product/Process Reliability. IEEE Workshop Accelerated Stress Testing. Reliability (ASTR 2009). Proceedings on CD," October 7–9, 2009. Jersey City.
14. Lev M. Klyatis, Oleg I. Teskin, James W. Fulton (2000) Multi-Variate Weibull Model for Predicting System Reliability from Testing Results of the Components. Annual Reliability and Maintainability Symposium Proceedings. Los Angeles, California, USA.
15. Lev M. Klyatis (1997) Prediction of Reliability and Spare Parts of Machinery. *ASQC's 51ˢᵗ Annual Quality Congress Proceedings*. Orlando, Fl, May 5–7, 1997.
16. Lev Klyatis. (2012) "About Trends in the Strategy of Development Accelerated Reliability and Durability Testing Technology." SAE 2012 World Congress. Paper 2012-01-0206.
17. Lev M. Klyatis (2011) "Why Current Types of Accelerated Stress Testing Cannot Help to Accurately Predict Reliability and Durability?" SAE 2011 World Congress. Paper 2011-01-0800. Reliability and Robust Design in Automotive Engineering SP-2306. Detroit, MI, April 12–14, 2011.
18. Lev Klyatis (2010) "Analysis of the Current Situation in reliability, Quality, and Durability, and the Reasons for this Situation. SAE 2010 World Congress. Paper 2010-01-0694. Reliability and Robust Design in Automotive Engineering SP-2272. Detroit, MI, April 12–15, 2010.
19. Lev Klyatis (2006) "A New Approach to Physical Simulation and Accelerated Reliability Testing in Avionics," Development Forum of Aerospace Testing Expo2006 North America, Anaheim, California, November 14–16, 2006
20. Lev Klyatis (1992) Test Center "Testmash". Journal Automotive Industry. 9/1992.
21. Whalley & Barley (1997) Technical Work in the Division of Labor Stalking the wily anomaly. University of Texas.
22. Kunda Gideon (2009) Engineering Culture. Temple University Press.
23. Klyatis Lev M. (2012) Accelerated Reliability and Durability Testing Technology. Wiley.
24. Klyatis Lev, Klyatis Eugene (2006) Accelerated Quality and Reliability Solutions. Elsevier. UK.
25. Klyatis, L. (2012) "About Trends in the Strategy of Development Accelerated Reliability and Durability Testing Technology," SAE Technical Paper 2012-01-0206, 2012, https://doi.org/10.4271/2012-
26. Klyatis, L. (2011) "Why Current Types of Accelerated Stress Testing Cannot Help to Accurately Predict Reliability and Durability?," SAE Technical Paper 2011-01-0800, 2011, https://doi.org/10.4271/2011-01-0800.
27. Klyatis, L. (2010) "Analysis of the Current Situation in Reliability, Quality, and Durability, and the Reasons for this Situation," SAE Technical Paper 2010-01-0694, 2010, https://doi.org/10.4271/2010-01-0694.

28. Klyatis Lev (2009) Accelerated Reliability Testing as a Key Factor for Accelerated Development of Product/Process Reliability. IEEE Workshop Accelerated Stress Testing. Reliability. ASTR 2009. Proceedings on CD. October 7–9, 2009.
29. Klyatis Lev (2006) A New Approach to Physical Simulation and Accelerated Reliability Testing. In Avionics. Development Forum. Aerospace Testing Expo2006 North America. Anaheim, November 14–16, 2006.
30. Bailey Bill D. and Alter Howard (2014) No Weak Links. Use lean and quality tools to strengthen global supply chain performance. Quality Progress. June 2014.
31. Harrington James and Vochl Frank (2005) The Future of Quality. In fifteen years, will you recognize your organization? Quality Digest. December 2005.
32. Paton Scot M.. Juran (2005): A Lifetime of Quality. An exclusive interview with a quality legend. Quality Digest. December 2005.
33. Carlson Carl S. (2014) Which FMEA Mistakes are You Making? Quality Progress. September 2014.
34. U.S. Department of Defense (1949) Procedures for Performing a Failure Mode and Effects and Criticality Analysis.
35. Sharma Manoj and Sharma Varday (2015) Discovering the right path. Quality Progress. June 2015.

Chapter 4
Accelerated Reliability and Durability Testing Technology as Second Key Factor for Successful Prediction of Product Efficiency

4.1 Introduction

Accelerated testing relates mostly to laboratory testing, proving ground testing, intensive field testing, and most experimental research. It is termed accelerated testing, because it is used to obtain results quicker than would be obtained by normal field testing. This is not a new concept. Accelerated testing was actually developed hundreds of years ago. While there are many different approaches to accelerated testing, the effectiveness of any approach depends on how <u>accurately</u> its testing conditions simulate real-world conditions. The accuracy of this simulation during design and manufacturing influences the number of recalls, the degree of reliability, durability, safety, maintainability, and other technical and economic components of efficiency that will be experienced over a product's life time.

As discussed in Refs. [1] and [2], accelerated reliability and durability testing (ART/ADT) is based on simulation of real-world conditions better than other methods of accelerated testing. Therefore, it offers more accurate initial information for successful efficiency predictions and provides the greatest opportunity for improving technical progress and advancing technology in society.

It is important to remember that ART/ADT does not relate to the traditional accelerated life testing (ALT) methodology that does not help successfully predict any product's efficiency during service life.

Supplementary Information: The online version contains supplementary material available at https://doi.org/10.1007/978-3-031-16655-6_4.

L. M. Klyatis, *Prediction Technologies for Improving Engineering Product Efficiency*, https://doi.org/10.1007/978-3-031-16655-6_4

4.2   Current Status of Accelerated Testing

The current status of accelerated testing is detailed in the author's book [2]. Basic aspects will be summarized here.

4.2.1   Basic General Directions of Accelerated Testing Development

Both the literature and the industry practices provide many different varieties of accelerated testing (AT) as related to engineering products.

This author has classified them into five general directions of AT development which are depicted in Fig. 4.1. Let us briefly review this classification.

4.2.1.1   The First General Direction (Field Accelerated Testing)

The first general direction is field testing with more intensive usage than which is to be experienced in normal use. For example, an automobile is typically used only about 5–6 h per day. By using the same automobile 18–20 or more hours per day,

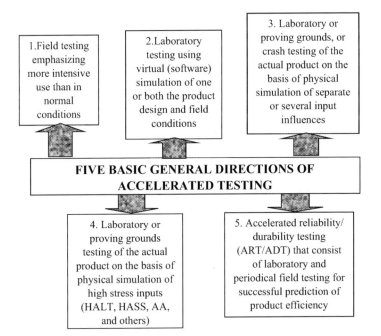

Fig. 4.1 Five general directions for accelerated testing development

the testing is assumed to represent true accelerated testing and should provide enhanced durability research into an automobile's parameters of interest. However, this approach also results in shorter nonoperating interval than those normally experienced (e.g., 5–6 h during accelerated use as opposed to 18–20 h in actual use).

This type of testing attempts to obtain quicker results than those gotten under normal field operating conditions. One such example is excerpted from a report of the US Department of Energy, INL/EXT 06-01262 [4] which stated:

> A total of four Honda Civic hybrid electric vehicles (HEVs) have entered fleet and accelerated reliability testing since May 2002 in two fleets in Arizona. Two of the vehicles were driven 25,000 miles each (fleet testing), and the other two were driven approximately 160,000 miles each (accelerated reliability testing). One HEV reached 161,000 miles in February 2005, and the other 164,000 miles in April 2005. These two vehicles will have their fuel efficiencies retested on dynamometers (with and without air conditioning), and their batteries will be capacity tested. Fact sheets and maintenance logs for these vehicles give detailed information such as miles driven, fuel economy, operations and maintenance requirements, operating costs, life-cycle costs, and any unique driving issues.... [4]

However, this type of field testing cannot be used for accurate reliability, durability, quality, maintainability, and other components of efficiency prediction.

There are several reasons for this, including:

1. The term "accelerated reliability testing" as used in the report does not correspond to an accurate definition of the term as stated in Ref. [1].
2. Many years of field testing of several specimens is required to obtain accurate initial information needed for successful quality, reliability, maintainability, and other components of efficiency prediction during a given period.
3. The methodology that can accomplish this objective and do it much faster with a lower cost can be found in Ref. [1].
4. Companies usually change their design and manufacturing practices every several years, and not always on a regular basis. In such situation the test results obtained from previous models may have only relative usefulness, which may not be directly applicable to current production models.

As will be shown later, field testing can only provide partial information for identifying problems related to an integrated system of quality, reliability, safety, and maintainability during the service life or warranty period of a product. For example, experience has shown that even after its field testing of its new models, Toyota (and other companies) still had problems in the areas of reliability and safety. These problems ultimately led to faster than predicted degradation, product failures, complaints, and recalls.

This accelerated field testing is conducted by professional drivers for short periods of time (maximum 2–3 years). Consequently, this testing paradigm cannot provide the necessary information for an accurate prediction of reliability, life cycle costs, safety, and maintenance requirements during the real service life, because it does not take into account the following interactions experienced during the service life of the vehicle:

- The corrosion process and other output parameters, as well as input influences that affect its service life
- The unreflected influences of real-world customers' driving habits and maintenance standards on the automotive reliability, because it was operated and maintained by professional drivers and mechanics during the above testing
- The influences of many other real-life situations such as variety of roads, driving purposes, etc.

Another example is that Mercedes-Benz calls similar testing as "durability testing." The test program for the new Mercedes-Benz C-Class was described as [5]:

> …For the real-life test work, that involved 280 vehicles expending a wide range of climatic and topographical conditions. Particularly significant testing was carried out in Finland, Germany, Dubai, and Namibia. The program included tough "Heide" endurance testing for newly developed cars, equivalent to 300,000 km (186,000 miles) of everyday driving by a typical Mercedes's customers. Every kilometer of this endurance test is around 150 times more intensive than normal driving on the road, according to Mercedes. Data gathered is used to control test rigs for chassis durability testing….

Of course, again this testing is improperly called "durability testing" for the same reasons as those listed in the previous example of what Toyota referred to as "accelerated reliability testing."

A similar situation existed with the Ford Otosan's 2007 durability testing [6]. In an article published about LMS Supporting Ford Otosan in Developing Accelerated Durability Testing in 2007, it was stated that "Ford Otosan and LMS engineers developed a compressed durability testing cycle for Ford's durability testing cycle for the Ford Otosan's new Cargo truck. LMS engineers performed dedicated data collection, applied extensive load data processing techniques, and developed a 6–8-week test track sequence and 4-week accelerated rig test scenario that matched the fatigue damage generated by 1.2 million km of road driving" [6].

4.2.1.2   The Second General Direction (Accelerated Testing Based on Computer/Software Simulation)

This direction involves virtual testing based on computer (software) simulation or analytical/statistical methods. A computer simulation is a computer program that attempts to simulate an abstract model of a particular system.

Computer simulations have become a useful part of the mathematical modeling of many systems in engineering, and provide insight into the operation of those systems.

Computer Simulation and Statistical Analysis
There are many companies that use computer simulation and statistical analysis. For example, Dynamic Research, Inc (DRI) [7] uses a wide variety of computer simulation and statistical analysis methods to perform research and development projects for their clients. These computer simulation methods include multi-body and finite element approaches for crash testing, and other types of research and testing.

Statistical analysis can be applied to simulation results as well as to their state, national, and international accident databases.

There are many approaches of computer simulation. For example, here is a sample of available software.

Computer Simulation Software
- LS-DYNA
- MADYMO
- ATB
- Nastran
- CarSim, TruckSim, and BikeSim

Statistical Analysis Software
- SPSS R
- Matlab

Commonly Used Databases
- FARS
- Hurt
- CPSC
- NASS/CDS
- NASS/GES
- NASS/PCDS (Pedestrian Crash Data System)
- MCCS (US Motorcycle Crash Causation Study)

Vehicle and Driver Vehicle Software Simulations (DRI) [7] has developed and applied on behalf of its clients a range of computer simulations for vehicle dynamics and control analysis.

Versions are available for a wide range of vehicles, including:

- Cars
- Trucks and utility vehicles
- Articulated vehicles
- Motorcycles
- All-terrain vehicles
- Buses
- Airplanes
- Space vehicles

DRI's applications also entail consideration of a range of different areas including:

- Driver-active control of vehicle motions (e.g., steering, throttle, braking in a variety of on-road and off-road tasks, maneuvers, and conditions, for predictive modeling of driver behavior and handling performance
- Human body active and passive response as related to vehicle control (e.g., human limb impedance, body-active control of small vehicles, etc.)

- Human body biomechanical response to impacts and large amplitude motions, for purposes of vehicle crash and rollover simulations
- Human injury potential models, probabilistic models based on measured biomechanical forces and motions, and expressed in normalized injury cost terms, as are useful in injury risk-benefit analyses
- Human comfort rating models, statistically derived empirical models of juries of human subjects, for predictive quantification of comfort due to measurable physical variables, in the areas of vehicle ride, handling noise and vibration

In Ref. [8] the company wrote: "Under relentless pressure to shorten development cycles and reduce costs, auto manufacturers are increasingly exploring ways to use analysis tools to perform meaningful virtual evaluations of vehicle designs early on, well before the availability of physical prototypes. The ability to obtain accurate predicted - or calculated – loads in this fashion augments the effectiveness of component physical testing, enables early identification and elimination of design flaws, reduces the need for rework and reliance on prototypes, and streamlines design validation."

The virtual proving ground method involves "driving" a vehicle model over a digitized road. But the company further admitted that, while having the advantage of relying entirely on virtual models, this approach yields inaccurate loads that are difficult to validate through physical testing. A semi-analytical method that uses spindle loads acquired from an existing vehicle to excite the vehicle model delivers better results than the virtual proving ground. But still it establishes boundary conditions not entirely appropriate for specific vehicle model, and calculated loads still lack accuracy and are difficult to validate.

A *company MTS Systems Inc.* [8] has developed a program called virtual testing that overcomes these problems by integrating a model of an actual physical test system into the simulation to excite the vehicle model.

Virtual testing is the simulation of a physical test, using finite element analysis tools, multi-body dynamic analysis tools, and RPC iteration techniques to derive accurate loads, motion, and damage information of a vehicle system early in the development process. MTS, Inc. has written claiming that there are several advantages of this approach [8]. First is that because it is easier to model the constraints of a physical test system than proving ground surfaces or tires. Second, virtual testing leverages an arsenal of proven, well-established physical testing tools and techniques, which have demonstrated utility in the analysis realm. And third, the incorporation of a modeled physical test system greatly streamlines the validation of results through subsequent physical testing. Spanning the analytical and physical test disciplines, virtual testing requires advanced knowledge of both CAE tools and physical testing, the development of a process to link MTS's RPC Pro software and analysis models, and preferably some degree of virtual testing exposure and experience.

MTS Systems Corporation has focused on refining the virtual testing approach, conducting a variety of demonstration projects with customers like Hyundai Motor Company (HMC) and Thermo King to validate their various methodologies. These projects included virtual tests of full vehicles and subsystems. The basic methodology from these experiences comprises the following steps:

1. Connect test rig models with specimen models.
2. Couple the models with RPC Pro.
3. Reproduce road load data on the virtual test rig.
4. Extract calculated loads.
5. Create the physical component/subsystem test.

A recent demonstration project with the SAIC Motor Corporation Limited provides an example of this methodology in practice:

1. Connect test rig models with specimen models.

At the outset of the project, MTS built a collection of test rig models in ADAMS, Simulink, and ADAMS Simulink co-simulation formats. The MTS test rigs that were modeled included a variety of Model 329 Spindle-coupled Road Simulators, a Model 353.20 Multi-Axial Simulation Table (MAST) System, and a Test Line component test system. Select test rig components were also modeled, including a FlexTest digital controller, MTS actuators and servo-valves, and a transformation that converts actuator displacement, acceleration, and force into degrees of freedom (DOF) displacement, acceleration, and force.

Due to project time constraints and the relatively slow simulation speeds of the more complex co-simulation models, the ADAMS/Car 329 model was coupled with an SAIC full vehicle, front suspension, and rear suspension models and ultimately used for the bulk of the virtual testing [8].

2. Couple the models with RPC Pro [8].

A virtual test server was developed to connect RPC Pro software and the ADAMS/Car 329 model. During testing, RPC Pro employs this virtual server to send drive files to the ADAMS model, initiate the ADAMS simulation, and copy the response file from the ADAMS folder to the RPC Pro working directory. Additionally, a Matlab Interface Tool already existing in RPC Pro was used to couple RPC Pro with the co-simulation models.

3. Reproduce road load data on the virtual test rig.

The RPC iteration technique was then used to reproduce road load (spindle force) data collected from the proving ground for a variety of select events and maneuvers. Interestingly, initial RPC iterations on the virtual 329 rigs exhibited no convergence, suggesting flaws in the SAIC vehicle short order. With the model improved, the RPC iterations finally converged, showing correlation between desired and achieved signals and RMS errors across all channels for both full and partial vehicle virtual simulations.

Using software simulation has positive aspects in comparison with physical simulation such as being cheaper and obviating driver professionals. The negative aspect of this direction is that the algorithms for this simulation currently cannot take into account every component of real-world conditions and, therefore, cannot help to obtain sufficient accurate initial information for successful prediction of the product efficiency.

These methods have become more popular over the last dozen years, especially by academia, because it is cheaper and simpler. But virtual simulation for the goal of successful, comprehensive prediction does not simulate the whole complex of real-world conditions, because it cannot reflect the interconnection between all the components of any product, especially complicated product.

4.2.1.3   The Third General Direction (Laboratory, or Proving Ground, or Crash Testing with Physical Simulation of Field Conditions)

This direction is an old approach of accelerated testing wherein the vehicle is subjected to simulations of field separate influences via special equipment (vibration test equipment, test chambers, or proving ground, or others) that generated product loadings much higher than that encountered in actual usage. This direction of testing does not provide accurate physical simulation of full range of many field input influences and many components of safety issues and human factor influences.

Unfortunately, too often such testing wrongly simulates field input influences separately such as temperature, humidity, radiation, pollution, or only limited combinations of field input influences. In such cases, this type of testing cannot offer accurate initial information for providing successful efficiency prediction. As a result, the real level of quality, reliability, durability, or other efficiency components is usually different than predicted.

For example, as described in Ref. [9], critical vibration testing requires accurate, reliable controls with advanced safety features to protect valuable hardware. Random vibration testing is used to verify strength and structural life by introducing random vibration through the mechanical interface. Random tests are typically performed in the frequency range of 10–2000 Hz.

Sinusoidal vibration testing includes testing from low levels to verify natural low-frequency responses through higher frequency levels to verify the strength of structures. Responses are monitored, and input forces are reduced or limited as necessary to ensure that the target responses or member loads are not exceeded. This testing, as well as others detailed below, is not based on accurate simulation of field conditions that have a more complicated random character.

An accelerated test plan considering an economic approach was described in [10]. It was introduced as a general framework to develop plans of accelerated testing with a specific objective such as controlling the cost of the testing.

The test plans were developed assuming considering a prior knowledge of reliability, including the reliability function and its scale and shape parameters, and the appropriate model to characterize the accelerated life. This information is used in Bayesian inference to optimize the test plan. In this analysis the prior knowledge is used to reduce the uncertainty of the reliability of the new product.

This methodology consists of defining an accelerated testing plan while considering an objective function based on economic value. It uses a Bayesian inference that statistically isolates the uncertainty of the parameters to obtain a robust, testing plan that meets desired cost parameters. In considering the above method, it should be remembered that one can use this approach for fatigue testing, but it is not appropriate for providing reliability testing and durability testing that is much more complicated process than fatigue testing and analysis.

The authors of Ref. [10] attempted to develop their test plan by extending their approach to include a theoretical formulation of the various degrees of freedom related to the test's parameters. To complete this development, the authors improved the algorithm of optimization of their test plan. The proposed method was illustrated by a numerical example based on a sample problem.

Finally, the authors introduced a general framework to provide an accelerated test plan with the inclusion of a cost objective. The cost objective function is developed in a theoretical formulation with the test plan parameters. Then, this framework is compared with the results obtained from a genetic algorithm. As the authors wrote, frequently, the genetic algorithm is a discrete stochastic process that can be considered as a Markov process. They wrote also that several results can be derived from this process, thereby enabling an easy verification of the "optimization's" test plan.

During the early 1960s, a number of research organizations did extensive bench testing on small mechanical components such as instrument size ball bearing, slip rings, and gears in a vacuum environment. In this testing, temperature is one of the many factors used for accelerating failures.

One obvious effect of increasing temperature is to increase the pressure in the vacuum chamber especially if liquid lubricants are being used, or if the test pieces are expelling gases. Temperature changes also affect the viscosity of a liquid lubricant and thus simulate the performance of a lighter oil, but this requires very precise instrumentation to measure and control temperature levels. One example of such accelerated life testing was when a lead-lubricated bearing was run at 100 rpm with periodic slowdowns to 16 rpm so that the torque spectrum could be measured. The test was run for 1×108 resolutions, almost 700 days at 100 rpm without developing excessive noise from torque. That was the equivalent of more than six times the required life of the product. In theory, by carefully increasing the severity of the test conditions, it should be possible to hasten early failures without altering the actual mode of the failure.

Horiba & MIRA presents [11] some of their wealth of vehicle accelerated test and development. All their activities are fully risk assessed. As part of the test activities, regular vehicle inspections are undertaken, periodic vehicle measurements made, daily checklists completed, and servicing and updates completed, and instrumentation is fitted and logged and vehicle audits are completed. All these can be uploaded to a secure web portal for instant customer access.

Of course, proving ground testing cannot account for every eventuality encountered when the product is used by the customer.

In 1889, the Swedish chemist Svante Arrhenius developed a mathematical model for the influence of temperature on chemical reactions. Because chemical reactions are responsible for some failures, the Arrhenius equation has been adopted to model the acceleration of testing.

Until now many professionals used different product and types of testing the Arrhenius equation and assumed an exponential distribution for time-to-failure distribution. Even many engineering books continue to use this exponential distribution for different types of testing. But in [12] it was written that using this distribution in engineering, including in consensus standards, "...have been proven inaccurate, misleading, and damaging to cost-effective and reliable design, manufacture, testing, and support."

This author demonstrated in his publications that this mathematical model is very old and the equation is inaccurate when used to simulate real-life conditions, because those conditions are more complicated and consist of many more input influences that only temperature. It is his contention that this model should not be used in reliability/durability testing of automotive, aerospace, aircraft, and other complicated engineering products without accounting for the interaction of different input influences, as well as human factors and safety problems. One can read this in more detail in Chap. 5.

Crash Testing

Usually, crash testing is considered separately, because it relates to second, third, and forth general directions of accelerated testing development. But mostly this type of testing relates to laboratory testing.

Crash testing is a form of testing usually performed to ensure safe design standards in crash worthiness and crash compatibility for various modes of transportation or related systems and components [13].

Crash testing is used to help reduce losses such as deaths, injuries, and property damage from vehicle crashes. In the United States, one organization responsible for ensuring the correct undertaking of this task is the Insurance Institute for Highway Safety (IIHS), a nonprofit, independent scientific body and educational organization. Another organization is the Highway Loss Data Institute, which supports the mission of the IIHS. Auto insurers support both organizations.

As an outcome of these tests, airbags are one of the most important safety devices [14].

They provide crucial cushioning for people during a crash. The devices are normally hidden from view but inflate instantly when a crash begins. Frontal airbags have been required in all new passenger vehicles since the 1999 model year. Side airbags aren't specifically mandated, but nearly all manufacturers include them as standard equipment in order to meet federal side protection requirements [14].

Some vehicles now have rear-window curtain airbags to protect people in back seats or front-center airbags to keep drivers and front-seat passengers from hitting each other in a crash. There are also inflatable safety belts aimed at reducing rear-seat injuries.

There are multiple kinds of crash tests for vehicle safety undertaken to provide the necessary information and guidance to vehicle owners. Examples of crash tests are frontal-impact tests, an offset test, side-impact tests, a rollover test, and roadside hardware crash tests.

Frontal impacts are test impacts against a solid concrete wall at a precise speed. SUVs are singled out from frontal-impact tests. An offset test requires only a portion of the front of the car to impact the barrier or a vehicle. An offset test is important since impact forces in this type of test remain the same as those with the frontal-impact test, but a small portion of the car is necessary to absorb all the force of the impact.

Side-impact tests are also important since side-impact accidents in vehicles result in a high fatality rate. This happens because cars usually do not have a significant crumple zone to cushion all the impact forces before the occupant is injured.

A rollover test verifies the car's ability to support itself, especially from the pillars supporting the poof, during a dynamic impact.

Roadside hardware crash tests ensure that crash barriers and crash cushions protect the passengers of the vehicle from roadside hazards. This kind of crash test also makes sure that some appurtenances such as sign posts, guardrails, and light poles may not be dangerous to vehicle occupants.

Many different crash test programs are practiced around the world, and all are dedicated to providing vehicle owners and drivers with data regarding the safety performance of new and used vehicles. These test programs provide safety performance based on field crash data [14]. They have resulted in manufacturers providing car airbags for passengers is that they provide an additional level of protection in the event of a car accident. This added protection can be the difference in some circumstances between life and death.

Risks of Airbag Deployment Versus Risk of Injury or Death

Although some question the overall safety of vehicle airbags because of instances of chest injuries and other impact injuries due to the nature of an airbag's deployment, these risks are lower than the risk of serious injury or death that may occur if the passenger is unprotected.

Disadvantages of Having Passenger Car Airbags

Unfortunately, these safety tools can have some significant liabilities:

- The biggest negative to airbags is that, though they are designed to protect, deploying airbags can injure occupants in some situations. The impact of a deploying airbag can injure a passenger who is improperly positioned. Deployment injuries can be most harmful to children and infants.
- Types of injuries from airbags include chest injuries, concussions, and whiplash.
- Resetting airbags after airbags have been deployed; they may be difficult to reposition for the next deployment. Car owners may spend a substantial amount of money at a shop getting new airbags after a deployment if there is only one occupant in the car, as a multiple airbag deployment can be very expensive.

The earliest automotive crash tests were morbid, often messy affairs. In the 1930s, in order to simulate the effects of high-speed collisions on drivers, researchers took human cadavers and subjected them to head-on crashes and vehicle rollovers. Anthropomorphic test devices (ATDs) emerged next. Those were the gleaming, faceless crash test mannequins shown in TV ads. These crash test devices were then followed by computer simulations.

While early ATDs were able to provide data on around 20 points of the body using a mesh of accelerometers, force sensors, and strain gauges, today's crash test computer simulations monitor the effects of a huge variety of crashes to a far greater degree of accuracy across a wide range of human body types, ages, and driving positions.

Their main benefit is accountability of all crash effects [15]. In traditional crash tests, researchers must anticipate where to best place the cameras inside the vehicles to monitor what happens during impact.

With modern methods, these placements have been improved and can show crash effects far better than earlier models.

Toyota has been at the leading edge in the crash testing field since the late 1990s when, in cooperation with Toyota Central R&D Labs [15], it began development of its inaugural virtual crash test dummy, which it called Total Human Model for Safety or THUMS.

Over the years THUMS has been involved in thousands of virtual crashes while slowly gaining new abilities. In 2004 it could monitor crash effects on facial bone structure. Then, the 2006 version through evolution added precise modeling of the brain in order to see how it might be affected in a variety of crash scenarios [18, 19].

The fourth THUMS generation added detailed modeling of internal organs, and the current version launched in 2015 added a comprehensive muscle model. The result is a digital model that contains no fewer than 1.8 million elements that combine to reproduce the human form, from precise bone strength to the structure of organs, which can be used to evaluate injuries to both soft and bone tissues.

The THUMS is currently used in safety research by other major carmakers, including Audi, Volvo, Renault, and Daimler, and numerous components suppliers. NASA has used this approach in the design process for Orion, a spacecraft that could lead the way for taking humans to Mars [17].

Virtual crash test dummies are being used in new and expanded roles in testing. But while these virtual crash dummies may seem to have supplanted their physical testing rivals, they cannot account for all the variances associated with advanced physical accelerated testing. This is especially true for product successful efficiency and it's component prediction [20].

Still, while crash test dummies have been the subject of public service announcements, cartoons, parodies, and even the name of a band, real crash test dummies are true lifesavers as an integral part of product crash testing. Sadly, while automotive-related fatality rates are declining, car crashes are still one of the leading causes of death and injury in the United States and other countries. A significant number of these crashes can be attributed to mechanical or electrical failures that might have been prevented through better pre-production testing.

Another leader in the field of safety research for automotive applications is Mercedes-Benz. In the end of the 1950s, Mercedes-Benz began practical testing for safety research purposes [17]. Initially, individual components were tested but shortly thereafter entire systems were crash tested.

Later impact tests conducted on newer components were also tested for newer components such as the seat belt system that became available in 1958. From 1973 onward crash tests became possible at the new test center in Sindelfingen, Germany. On the 65-m acceleration track, a linear motor producing a thrust force of 53,000 Newton accurately pulled the automobiles into a 1000-ton barrier, which rested on a very sensitive force measuring platform. This crash test facility was thoroughly renovated in 1998 with the creation of the Mercedes-Benz Technology Center (MTC). At an expense of 2.3 million euros, the facility was refurbished with state-of-the-art technology. The length of the acceleration track was increased to 95 m; as a result, all types of crash test variants were now possible at the facility. In particular, this included offset crashes, in which only part of the frontal width of the vehicle impacts the obstacle and which in reality occur much more frequently than head-on vehicle collisions. The test sequences are no longer recorded by high-speed film camera, but by video technology that has a very high frame rate so that the crash tests can be evaluated in extremely slow motion.

The facility was also equipped with a roof during the renovation so that passenger car and commercial vehicle testing can now be performed regardless of the weather.

From the beginning of crash testing at the company, not only vehicles were used to assess the effect of crashes, but measuring instruments in the head and chest of dummies provided information on the loads acting on the driver during an accident. Sandbags and mannequins initially took the place of the front passenger; soon dummies were also collecting crash test data for the front passenger seat and rear bench seat. Increasing computer capabilities then allowed dummies to be replaced by mathematically simulated multi-body systems. The first digital crash computations with overall vehicle models were performed for the E-Class of the 124 series. Consequently, the head-on collision is increasingly being replaced by offset collision testing scenarios.

In 1992, an offset frontal collision was performed against a deformable barrier for the first time, providing results that corresponded even more closely to the behavior of a vehicle in a real-world accident. One outcome of this testing was the development of a more sophisticated deformable barrier for this type of experiment in Europe; its design was decisively influenced by test results from the Mercedes-Benz Safety Center in Sindelfingen.

The specifications called for a lower risk of passenger injury in the following cases:

- A head-on impact
- Against a fixed barrier at 80 km/h, a head-on impact against another vehicle at 120 km/h, a side impact by
- Another vehicle at 50 km/h, a rear-end collision by another vehicle at 120 km/h, and a rollover

A helicopter crash test program conducted by NASA, the US Navy, the US Army, and the FAA hopes to shedding new light on seat safety and lightweight composites while also exploring the value of new testing methods [20]. Engineers at NASA Langley's landing and Impact Research Facility (LangD1R), in Hampton, Virginia, sought to improve the crashworthiness of seats and seatbelts, as well as to gather relevant data on the odds of surviving a helicopter crash.

For this program, the US Navy provided the CH-46E Sea Knight helicopter fuselage, complete with seats, which were fitted out with 15 "occupants"—consisting of 13 instrumented crash test dummies and 2 instrumented mannequins. The Navy also contributed five of these crash test dummies, one mannequin and other equipment, while the Army provided a and a crash test dummy that was placed in a position representing of a patient in a medical evacuation litter. The FAA provided a side-facing specialized crash test dummy and part of the data acquisition system.

NASA Langley added six of its own dummies, as well as being the leader in technical expertise and provided the use of its own specialized facility, known as the Gantry. Engineers then used cable to hoist the helicopter fuselage into the air and swinging it above the ground like a pendulum. It was traveling at a speed of 30 mph when pyrotechnic devices separated the cables, sending the fuselage smashing into the soil below. The test article was fitted with 350 sensors to capture data on airframe accelerations and crash and test dummy loads. Over 40 high-speed and high-definition cameras recorded onboard and external movements.

Researchers also made use of a new photographic method to help analyze the data collected from the crash test. Called "full-field photogrammetry," it saw the helicopter fuselage which had been stripped of its usual coat of naval gray paint in favor of camera catching paint scheme. "We painted more than 8,000 dots on the side of the test article to measure global and local deformation on the fuselage skin," explained Annett a leading expert in testing using high-speed camera filming. At 500 images per second, this was used to track each dot, ensuring researchers were able to plot and "see" exactly how the fuselage behaved under crash loads.

Incidentally, a similar approach was used by the author of this book in the 1980s for automobile's accelerated reliability testing. This experience was published by Wiley in the author's book *Reliability Prediction and Testing Textbook*, as well as in other of his publications.

The first quadricycle crash tests by Euro NCAP were performed in 2014 with three electric and one petrol model. These vehicles were type-approved and met the minimal safety requirements [23] released by Euro NCAP at the time of the tests read. "All of the quadricycles tested showed critical safety problems, although some fared better than others in the front or side impact tests."

The vast majority of the data from the quadricycle tests comes from the dummies, and as there are fewer of them than in a normal car crash test, fewer cameras are needed.

Let us discuss one more approach to crash testing. The Center for Advanced Product Evaluation (CAPE) in Westfield, Indiana (United States), is a unit of the advanced vehicular safety systems manufacturer (MMI). The center designs and builds test rigs that help to determine whether there is survivable space inside the vehicle, and whether the vehicle's body-to-frame monitoring system is sufficient to withstand a rollover incident [24]. The testing performed at CAPE is typically designed to prove that the manufactured vehicle complies with standards set by different organizations.

CAPE completed development of a test rig that can provide vehicle OEMs with roof crush testing up to 100 tons. It can also test off-road vehicle roll cages and race car chassis. The rig uses four hydraulic actuators (cylinders), mounted at the four corners of a heavy-gauge pressure plate and controlled as four separate motion axes. At the core of this system is an eight-axis RMC150 electrohydraulic motion controller manufactured by Delta Computer Systems.

The typical compression cycle works as follows [24]. The hydraulic pump is turned on and the transducers are initialized to zero values. Then, the four compression cylinders are set up to be geared together, and the system is given a command to move the steel pressure plate up and out of the way.

The vehicle cap/body is placed in the rig, and the pressure plate is lowered until it reaches a position that is just above the cab but not touching it. The command is then given to preload the rig to 500 pounds, followed by the command to apply the full load, a process that takes between 1 and 5 min.

The system is allowed to rest under load for 30 s, and then it is unloaded to zero pounds on the load cells. Finally, the pressure plate is moved completely off the cab, and the test data is downloaded from the motion controller to the network drive over the RMC's Ethernet interface. Programming the motion steps was done using PMCTools software, provided with Delta's motion controllers. It enables programming the controllers using high-level commands, such as the virtual gearing arrangement. During testing operations, the Delta motion controller in the CAPE test rig performs the data acquisition and maintains all the test data internally. The PMCTools software can handle test system operator interface functions and data transfer to an attached PC.

Unfortunately, all of these current types of crash testing generally do not take into account human factors that are one of the basic reasons for real-life crashes. This author's books [20] and [2] cover these human factors, and how to take them into account during crash testing. Among these factors are lighting problems on the road, especially in big cities. The above books also considered what kind of human factors one has to take into account during research, design, and testing for successful prediction of product efficiency.

But the fundamental problem with current crash testing methodologies is how they are constrained to a separate type of testing. This denies the complexities of real-world situation and therefore is not successful for obtaining accurate information for successful efficiency prediction. These tests can examine the specific issues discussed above well enough, but until they are incorporated into broader scenarios that reflect the issues faced by users, their useful applications are very limited.

4.2.1.4   The Fourth General Direction of Accelerated Testing

In the literature we also often see that professionals are using HALT and HASS for the laboratory or proving ground testing, especially for accelerated reliability testing (ART) or accelerated durability testing (ADT). However, this also is not appropriate, because as Gregg Hobbs, the author and inventor of HALT (highly accelerated life testing) and HASS (highly accelerated stress screening), wrote in his book [23]:

> Design engineers involved in quality assurance and students of reliability engineering will benefit from this unique resource detailing the technical aspects of accelerated reliability engineering. Features include:
>
> - Coverage of the physics of failure and useful testing equipment enabling those new to the area to grasp the concepts behind HALT and HASS
> - Overview of the HALT technique demonstrating how to find design and process defects quickly using accelerated stress methodology during the design phase of the project
> - Examination of detection screens and modulated excitation used to detect flaws exposed in HALT
> - Description of how to set up a HASS profile and how to minimize costs whilst retaining efficiency
> - Applications of HALT and HASS and analysis of common mistakes highlighting the pitfalls to avoid when implementing the methods.

So, we can see that Gregg Hobbs did not consider his approach to be a means for accurate simulation of the real-life conditions and their interactions necessary for ART or ADT. Rather he intended it for other general directions of accelerated testing.

Unfortunately, many professionals wrongly used his approach as a method for any type of accelerated testing development. This is wrong, because HALT and HAAS do not depict, nor intended to provide accurate simulation of real-world conditions, including the interconnection of many different aspects of these conditions. Therefore, these testing results are different from real-world results and cannot be used for the successful prediction of product efficiency.

4.2.1.5   The Fifth General Direction (Accelerated Reliability/ Durability Testing)

Figure 2.2 depicts why accelerated stress testing (mostly related to ALT) cannot generate accurate initial information for efficiency successful prediction.

Accelerated reliability testing (ART) and accelerated durability testing (ADT) [20] are the most useful ways in the development of accelerated testing in all areas of engineering. They provide the greatest opportunity for improving technical progress and advancing technology in society, because they include accurate simulation of normal field conditions, as well as the interconnection of their components.

It is important to remember that this direction does not relate to accelerated life testing (ALT), which includes basically the traditional approaches to testing that cannot help successfully predict any product's efficiency during its service life. Instead, the author has developed a new direction of accelerated testing of engineering products, which he has called "accelerated reliability and durability testing" (ART/ADT) technology.

This direction has different approach from traditional accelerated life testing, because ART/ADT in comparison with ALT, has much more robust testing capabilities than those employed in ALT. Therefore, using ART/ADT [1, 2] provides accurate initial information for the product's successful efficiency prediction during any given time. ALT cannot offer this advantage. As this author demonstrated in many technical papers presented at the SAE 2002–2022 World Congresses in Detroit, ASQ 1997–2006 Quality Congresses, IEEE 1998–2000 and 2009 Workshops in Accelerated Stress Testing, RAMS (Reliability and Maintainability Symposiums) in 1999–2005, and ASAE (American Society Agricultural Engineers) in 1995–2003, this direction is particularly useful for any new product's successful efficiency prediction.

The following are the basic concepts of this new accelerated testing direction:

• Based on accurate simulation of the field conditions (quantity and quality of influences, character of change during the time of each influence, the dynamic of changes of all influences, as well as their interconnection with safety and human factors).
• ART/ADT consists of two basic components: (A) laboratory testing utilizing the simultaneous combination of each group of field influences (see Fig. 4.8) and (B) coupled with periodic (at regular intervals after the given hours of laboratory testing) field testing.
• Conducting simulation testing 24 h a day, every day, but not including:

  – Idle time (breaks or interruptions).
  – Time operating at minimum loading does not contribute failure of the product.

• Utilizing a complex system for modeling each of the interacting types of field influences. For example, pollution is a complex system that consists of chemical air pollution + mechanical (dust) air pollution. Both types of these pollutions must be simulated simultaneously for accurate predictions.
• Simulate a whole range of each type of field influences with their characteristics. For example, when simulating temperature, one must simulate the whole range of temperatures from the minimum to the maximum, the rate of temperature change, and the speed of temperature change.
• Maintaining a proper balance between field and laboratory conditions.
• Correcting (if necessary) the simulation system (methods and equipment) after comparison of the degradation and failures in the field and during ART/ADT.
• Using the physics-of-degradation process for an accurate simulation of the field conditions.
• Treating the system as interconnected by using system of systems approach.
• Considering how the interactions of test subject components act within the system.
• Reproducing the complete range of field operational schedules and maintenance or repair actions.
• ART/ADT technology is also based on the accurate simulation of all the interacting components (units and details) of the whole vehicle. Figure 3.12 in Ref. [20] visually depicts these interactions.

When ART/ADT is properly performed, it accounts for the interactions and accurate simulation of the different components under field conditions, including the whole vehicle or its units and details (depending on the test subject).

Figure 4.2 depicts ART/ADT as a higher level of accelerated testing, which is based on an accurate simulation of field conditions that leads to a successful prediction of a product efficiency with all associated components.

4.2.2   Other Testing Approaches

There are many other used testing approaches. They are well known and, therefore, considered briefly here. For example, the following six approaches provide a broad range of such approaches offering different specifics and limitations:

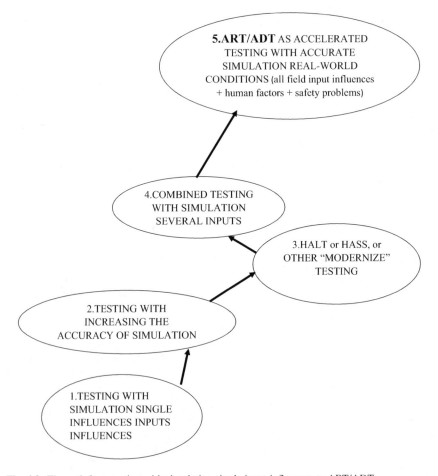

Fig. 4.2 The path from testing with simulation single input influences to ART/ADT

1. Time compression approach. The time compression approach simply operates the item more than it runs during normal use.
2. Build a stress to life relationship approach. The ALT approach is to apply higher than expected stresses (such as temperature, voltage, etc.) to the product in such a way that the failure mechanism's progression of interest is accelerated. This approach may lead to significant differences between the accelerated testing results and the field results; it can result in unsuccessful prediction of product efficiency and its components (quality, reliability, durability, maintainability, etc.) during the product's usage time.
3. Given acceleration model approach. This approach utilizes one or several of the many previously discussed models and equations to save time and cost of developing the testing models, and hence, the reduction of time from design to market. But this approach is not guarantee obtaining accurate information for successful prediction of the product efficiency.
4. Step-stress approach. This ALT approach should only be used in situations where the damage leading to failure accumulates proportionally to the application of stress, i.e., higher stress results in greater damage.
5. Shock testing approach. The shock testing approach was developed from the need to test the ruggedness of new products during the development process. The first shock pulses were generated by dropping products onto various impact mediums (e.g., sand) to reproduce shock inputs. These early testing methods were challenged by poor repeatability and consistency issues and were generally limited to a pass/fail type of test.
6. Degradation approach. The degradation approach is based on the feature of some failure mechanisms to exhibit measurable deteriorations of performance of the product life cycle. It is also not appropriate to successful efficiency prediction.

The above six (as well as some other approaches) to testing are primarily based on a statistical (mathematical) basis, and, therefore, are only partly related to the physical or chemical essential characteristics of reliability, durability, maintainability, life cycle, and other components of the product's efficiency.

Engineers generally prefer using the statistical aspects of failures rather than the physical aspects that result in the failing mechanism. It must also always be remembered that degradation and failures of technical components are only a small part of the product's quality, reliability, durability, and maintainability, and of the economic aspects such as life cycle cost, profit, and other aspects of a product's efficiency. Too often these physical aspects that can lead to the success or failure of a product are overlooked or ignored. This is usually because of parochial interests or improper attempts at cost saving by professionals involved in engineering the product's design and manufacturing. One indicator of this is just how infrequently literature discusses the role of human factors in deterring and resolving quality and reliability problems. Even in the volumes of literature being disseminated on autonomous vehicles where human factors should be a leading factor in designing systems of control, you will find people designing these systems primarily from the viewpoint of system control and maintenance and cost reliability. Too often there is minimal emphasis or consideration of the human-induced variations, especially when the human inputs are not the ones predicted by the designers.

The figure below illustrates why most traditional approaches to accelerated testing lead to unsuccessful prediction of product efficiency, including reliability, durability, and safety prediction. Their development and use should connect directly to the speed of new technical development. It is properly equal when speed of technical progress corresponds to the speed of technical progress in testing. This provides high quality level for new product. But, as shown in Fig. 4.3, the factual speed of technical progress in testing during the time is much slower than speed of technical progress in design, research, manufacturing, and service of new technology and equipment.

Now, as many years ago, one sees the modeling of only 1–3 input influences from many in real field conditions, such as separate corrosion, vibration, proving ground testing, environmental testing with associated simulations several from many factors of real field conditions, etc. This is considered in detail in author's books [1–3].

Moreover, the technical progress in testing has diminished during the last 10–12 years, because physical testing is giving way to the more popular virtual testing, where simulating reflects only a part of real-world conditions. As a result, success in efficiency prediction is not increasing. Example from author's experience of

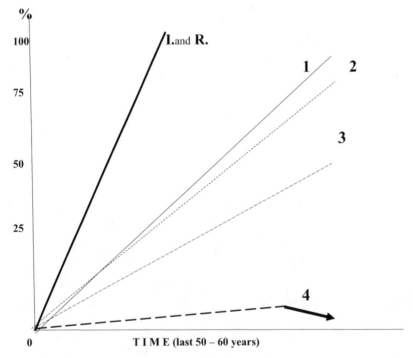

I and **R** – ideas and research; 1 – design, 2 – manufacturing, 3 – service, 4 – testingv

Fig. 4.3 Progress in different areas of engineering activity (last 50–60 years)

this situation can be seen now in Chap. 2 of this book. This situation is discussed in more detail in Ref. [20].

Now let us review the technical session of the SAE World Congresses in Detroit, which focused on accelerated testing. As we saw in this book earlier, reliability and durability testing is related to higher level of testing. The SAE 2012 World Congress organized a special technical session entitled "Trends in Development Accelerated Reliability and Durability Testing." It highlighted three presentations given there directly that were repeated at the SAE 2018 World Congress. Two presentations on the same topic were given in the technical session of the SAE 2021 Digital Summit. Only author's presentation there directly to reliability/durability testing at the SAE 2022 World Congress.

A similar situation was related to the Annual Quality Congress and Exposition (ASQ). The name of the congress was changed in 2017 to the ASQ World Conference on Quality and Improvement. The ASQ 2000 Annual Quality Congress and Exhibition included a special technical session "Accelerated Testing Is a Key Factor for Design and Reliability Problems Solving," which included four presentations directly in this area. Now we cannot find this subject being discussed in any subsequent ASQ World Conference programs.

In a paper presented at the SAE 2022 World Congress, the author reviewed articles published by Automotive Testing Technology International, November 2020, and Aerospace Testing International, 2020 SHOWCASE. The examples provided from these articles clearly showed how slowly the testing development is moving now [25]. Future discussion must emphasize the importance of defining customer's requirements for new testing technology and equipment development; engineers must take into account the following:

- For successful prediction of product efficiency, it is necessary to have a combination of sources for obtaining reliable information for calculating changing components and parameters of specific models efficiency during given time (warranty period, service life, or other time or work volume).
- Intensive field testing alone does not generate adequate information on corrosion, deformation, and other types of degradation during a product normal service life. Without going further with broader testing programs and obtaining this information, one cannot gain successful prediction of product efficiency.
- Accelerated laboratory or proving ground testing does not provide accurate simulation of real-world conditions. Therefore, results of these testing are erroneously different from real field results.
- Accurate physical simulation of real-world conditions requires a wide range of factors that are simultaneously interacted in the real field on the real product. It includes, besides of input influences, human factors and safety factors.
- Virtual simulation requires the same factors as physical simulation.
- Often top management do not want to invest money in new innovations, because they are not familiar with these innovations and corresponding benefits. Often, they delegate part of their responsibilities in the quality and reliability to other people, who do not have the authority to invest development money. This problem has been reviewed at length in previous books by the author.

For better understanding of current situation in accelerated testing, let us consider the following statement by the Mahle Powertrain Group:

> MPT (Mahle Powertrain Group) will open a second RDE (Real Driving Emissions) test chamber in Northampton (UK), providing space under a wide range of the climatic conditions, including the ability to simulate solar loading. The new test chamber will be equipped with a four-while-drive chassis dyno, a battery emulator and safety measures to enable testing of hydrogen-powered vehicles. The set up will include blast walls and the domed chamber roof for stay H_2 gas collection.
>
> The investment was prompted be heavy demand for the center's original RDE facility, which opened in 2018. Mahle research estimates 73% of passenger cars globally will still make use of an ICE as part of their propulsion system, and 50% will still employ ICE by 2040. MPT expects demand for the RDE test chambers to continue as ICE development refocuses on hybridized units. "Given that statistic, it therefore made reflect sense to have a second chamber with a battery emulator suitable for EV development", said David Pates, MPT's head of engineering.
>
> The latest MPT developments are part of the company's ongoing EV strategy, which additional phases planned to roll out in 2022 focusing on battery-development capabilities. The MPT goal is to address the entire battery-development process, from pack built, test simulation and analysis to real-world testing and strip-down of battery modules [24].

As we can see from the above, until now professionals wrongly used the term "real-world" simulation and testing, because they simulate in the test chambers only separate components and not including many of real-world factors (see Chap. 2). This also eliminates the possibility of considering how the test subject interacts with other components (units and details) of the whole system comprising a machinery. To improve this situation, one needs to understand that battery-development process interconnects with many other EV processes. If one wants to obtain correct laboratory testing results that reflect real-world results, one needs accurately to simulate all interacting real-world conditions, not just several separate processes. One can read this in author's books [1–3], and others.

Finally, it must be remembered that many of the statistical approaches to accelerated testing are primarily theoretical aspects of estimation, which do not consider the existing nonstationary random character of the processes that are experienced in real-life operation. If one will use the above approaches in the future, it will only generate further negative influences on the quality and efficiency of modern equipment and technology.

In the end, ART/ADT will have much more benefits for producers and users, because it eliminates most of the negative aspects of what in the present day is termed accelerated testing.

4.3   The Basic Methodology of Accelerated Reliability and Durability Testing

This methodology is detailed in author's book *Accelerated Reliability and Durability Testing Technology*, Refs. [1–3]. Here, however, are presented the basic components of this methodology.

Accelerated reliability and durability testing consists of two basic components (Fig. 4.4):

1. Laboratory testing, which offers the possibility for accelerated degradation of product through accurate simulation of real-world conditions and their intensivity
2. Periodical testing in natural real-world conditions, which helps to take into account factors that are impossibly evaluated in the laboratory with high accuracy

Accelerated laboratory testing includes the simultaneous combination of mechanical testing, multi-environmental testing, electrical testing, electronic and programming testing, crash testing, and others. Periodical testing in real-world conditions includes testing functional stability during usage time, evaluation usage cost changing during the time, testing safety during usage, operator's (driver's) and management reliability.

Step-by-Step Accelerated Reliability and Durability Testing (ART/ADT)
As was mentioned, one of the basic reasons why methods and equipment for ALT often offer low return benefits is that they often give inaccurate information not only for successful efficiency prediction but solving different reliability, safety, and other problems (evaluation, analysis, causes of product deterioration and failures, and others) as well. In the author's previous books were detailed how one can improve this ineffective approach. Among them are the important concern of providing correct problem definitions that ensure effective testing programs (see developed standard in Chap. 6) To ensure achieve of this definition goal, all of the author's books include a special chapter on Terms and Definitions.

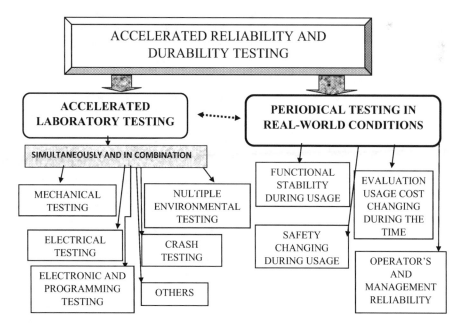

Fig. 4.4 Scheme of basic components of accelerated reliability/durability testing (ART/ADT)

Below one can see briefly the full step-by-step strategy, which can help to solve these problems.

It comprises the 11 basic steps that link with each other.

Step 1: Collection of the initial information from the field

This step is very important, because it helps to convey the kind of input influences that act on the specific product in the field. To determine how they act, select the input influences that must be simulated in the laboratory for accurate ART/ADT. Without this knowledge one cannot substantiate accurate ART/ADT parameters and regimes.

For this goal it is necessary to study:

- The full range of input influences which affect product reliability, durability, maintainability, etc. (Fig. 4.5) $X_1, X_2, X_3, \dots X_n$ (the range, character, speed, limit of value changing, etc. of temperature, air pollution, humidity, full complex of radiation, input voltage, air pressure, features of the road, physical-chemical-mechanical attributes that affect the product, etc.) under various field conditions wherever the product is employed, as it acts in the field.
- Input influences on the product in single steps and in simultaneous combinations including interaction, as they act in the field.

The system (test subject) hierarchy is that the complete product in real life consists of N units (subsystems) that act in series with interactions. The subsystems (units) also consist of K sub-sub-systems (details/components) that also act in series with interactions.

As a result, the system (complete product), its subsystems (units), and sub-sub-systems (details) provide their functions and the system works. If one wants to simulate real-life influences on the components accurately in the laboratory during ART/ADT, one must simulate accurately the full hierarchy of these connections and interactions as discussed below.

- Output variables $Y_1, Y_2, Y_3 \dots Y_m$ (vibration, corrosion, loading, tension, wear, resistance, output voltage, decrease of protection after deterioration, etc.) the

Fig. 4.5 Scheme of input influences and output variables

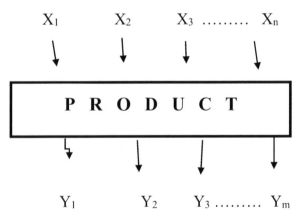

range, character, speed, and limit of value change under various field conditions including climatic areas. With a new product, these studies can be brief, using additional data from any prototypes. There will be variations to this list of variables depending on the facilities of the researchers, the goals of the research, conditions of experiments, the subject of the study, etc.

- Mechanism of degradation of the components or test subject (parameters of degradation, value of parameters of degradation, dynamic and statistical characteristics of these parameters which change during the usage time, etc.).
- Input influences and output parameters, data collection, and analysis. What types of input influences affect the degradation and failure process? If we are certain that any influences do not affect the product degradation (failure), they may be eliminated from the next review's consideration. But we must what they exactly. This possibility was described in detail by Parker P. [26] for electronic equipment.
- The distribution percentage of the test subjects among the different usage conditions. This is important, because most products are used in wide variety conditions and loads. These conditions change the values of each output parameter. Importantly values are needed for programming and understanding the ART/ADT results.
- The complete design of the test subject as a system which consists of subsystems (units) and sub-sub-systems (details) with their connections and interactions. One must remember that units and details are not working separately but in connection with each other. Understanding of this is especially important to suppliers.
- The work breakdown schedule tool to highlight this design and, thereby, highlight relationship among components.

Step 2: Analysis of the above information as a random process with evaluation of the statistical characteristics of the studied parameters

Most of the field input influences and output parameters have a probability (random) character, because they change during usage and the character of this change is random [1]. Therefore, one has to evaluate as a basic minimum the mean, standard deviation, correlation (normalized correlation) or power spectrum, as well as the distribution of input influences and output parameters. Figure 4.6 shows an example of ensembles of experimental normalized correlation and power spectrums of tension data registered by sensor 1 for the car's frame point for different field or operating conditions.

Step 3: Establishing concepts and statistical criteria for the physical simulation of the input influences on the product selected for ART/ADT

The field input influences must be accurately simulated in the laboratory with a limit of stresses to give a higher correlation between ART/ADT results and field results, and for an accurate prediction of field reliability, durability, maintainability, and other efficiency components after ART. In general, the most accurate physical simulation of the input influences processes occurs when each statistical characteristic $[\mu\ D\ \rho(\tau)\ S(\omega)]$ of all input influences differs from the field condition measurements by no more than 10%.

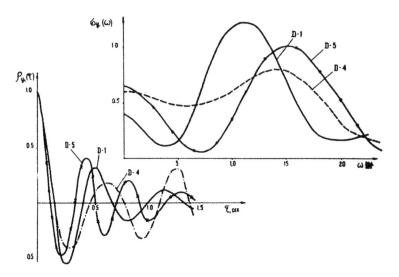

Fig. 4.6 Ensembles of normalized correlation $\rho_{\varphi}(\tau)$ and power spectrum $\sigma_{\varphi}(\omega)$ of frame tension data of the car's trailer (the author's experimental data)

In specific situations one has to calculate and use the statistical criteria for this goal of comparing the efficiency measured by an ART/ADT results to the field results (see Chap. 3), but all these methods have an acceleration coefficient (Fig. 4.7). The similarity of the degradation process in the field and in the ART/ADT will determine the practical limit of stress (acceleration coefficient) (see Fig. 4.8).

The ART/ADT process times can be decreased by minimizing idle time between operational time and reducing the influences that does not lead to test subject failures. This type of testing can be accelerated if the product is tested 24 h a day, every day, and it does not include idle time, or time with minimum loading. This approach is based on the principle of reproducing the complete range of operating conditions and maintaining the proportion between heavy and light loads. The author's experience shows that this has the following basic advantages:

- Maximum correlation between field and laboratory testing results.
- One hour of pure work performed by the product that faithfully reproduces its stress schedule is identical in its destructive effect to 1 h of pure work under normal operating conditions.
- There is no need to increase the pace of testing in terms of the size and proportion of stress in this approach.

This is accelerated testing, because the result of testing in the laboratory is faster than in the field by more than ten times. This method is especially useful for a product that works intermittently for short periods of a time.

Specialists who use this method know that reproducing a complete range of operating conditions is not easy, but gives a valid correlation between ART/ADT results and field results.

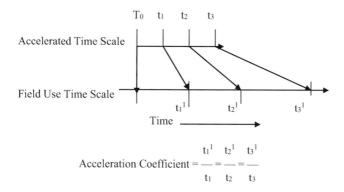

Acceleration Coefficient = $\dfrac{t_1{}^1}{t_1} = \dfrac{t_2{}^1}{t_2} = \dfrac{t_3{}^1}{t_3}$

Fig. 4.7 Accelerated coefficient describing

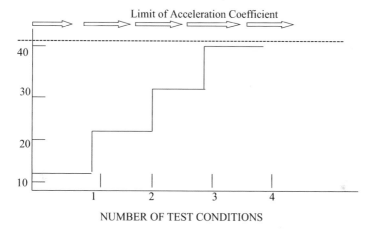

Fig. 4.8 Example of stress limit (acceleration coefficient)

To obtain the accelerated testing results, one can use the stress methods. However, the correlation between laboratory test results and field results will be less. More stress means less correlation. Less correlation means less accuracy of simulation and greater problems with accurately evaluating and predicting components of product efficiency (reliability, durability, etc.) in the field. Less correlation brings more problems with finding the reasons for degradation and failures (and eliminating them), accuracy of simulation, and greater problems with accurate evaluation and prediction about the product.

Therefore, in the case of accelerated coefficients, one must decide specifically for specific products which way will be more effective.

This selection depends on the testing goal. Therefore, if it is necessary to have an independent simulation, for example, for vibration testing of the product, one must know whether after this testing it is possible to evaluate accurately the reliability or durability of that product. The same is true with simulation the temperature conditions for temperature testing of the product. If one simulates only one parameter, then it is too limiting and does not reflect the real world.

Therefore, a decision must be made to select other parameters that are realistically comprehensive. Without proper choices being made beforehand, these types of testing cannot give accurate information about the reliability of the product under accelerated testing actions given minimum correlation between accelerated testing results and field testing results. One must always take care with a high-level range of acceleration, as well as a combination of the field influences and other aspects of ART/ADT.

Step 4: Development and use of the test equipment which simulates the field input influences on the actual product

There are special design effects on test equipment designs that result from analyzing the combination of field input influences on the product ($X_1 \ldots X_n$) and how they can be simulated in the lab.

This can allow the usage of either universal and specific test equipment depending on the testing circumstances. Universal test equipment can be used on many different types of the product. Usually, they are used for design and manufacturing problems by special companies. For example, all types of mobile product and parts of the stationary machinery in the field vibrate. Therefore, they need vibration testing and for this goal vibration equipment must be selected. The same is true for test chambers which simulate environmental influences, and combinations of multi-environmental influences with other types of influences.

The capability of test equipment is an important requirement factor as it exerts an influence on the level of accelerated testing results. To provide ART/ADT, one needs test chambers, where there is simulated a maximum number and types of field influences.

Each type of simulation input influences in the chamber must reflect real life as closely as possible. For example, a company needing test equipment can buy single axis and (or) multi-axis vibration test equipment (VTE). For mobile products single-axis VTE cannot simulate real-life vibration. The modern solution in this area is multi-axis VTE.

For example, vibration input can be generated in up to three or six degrees of freedom. It is 1–3 linear (vertical, lateral, and longitudinal) and from 1 to 3 angular rotational (pitch, roll, and yaw). This VTE has to provide random vibration as in real life. The user can control the motions of each actuator independently or in groups to determine the degree-of-freedom control.

Environmental accelerated testing (EAT) can also be used for real-life environmental simulation. The temperature/humidity equipment is only a part of environmental influence simulation in the laboratory. A higher-level EAT simulates combined temperature, humidity, pollution, radiation, air, gas, and/or water pressure. Equipment for measuring these influences of simulation can also be at different levels. For example, pollution simulation can be a mechanical (dust) simulation (low level) or a high-level simulation using a combination of mechanical (dust) and chemical (chemical conditions of the environment) simulation. High-level simulation leads to more accurate simulation of field conditions.

Many products (or group of similar products) have unique designs and need specific testing equipment that accurately simulates relevant field input influences that ensure their availability and reliability of use. For this simulation process, one needs specific test equipment.

To meet those special requirements, many industrial companies design, manufacture, and use their own specific test equipment. Both universal and specific equipment may be required to work as a simultaneous combination if one wants to obtain accurate initial information for successful prediction in the results of ART/ADT.

Step 5: Determining the number and types of test parameters for analysis during ART/ADT

The types of simulation methods depend on the specifics of product use and the limits of the facilities of the test company.

One cannot simulate in the laboratory all the various operating conditions and their input influences. The basic objective is to determine the minimum number and types of test parameters to enable comparison of the laboratory and field testing results that are sufficient for accurate efficiency predictions components (reliability, durability, maintainability, etc.).

Therefore, it is necessary to establish the partial (basic) area of each influence that can be introduced into all varieties of operating conditions to set a minimum acceptable test condition.

This is:

$$E > N,$$

where E is the number of field input influences $X_1 \ldots X_a$ and N is the number of simulated input influences $X^1_1 \ldots X^1_b$.

And allowable error simulation input influences $M_1(t)$:

$$M_1(t) = X_1(t) - X_1^1(t)$$

where $X_1(t)$ are input influences of the field and $X_1^1(t)$ are simulated input influences.

The author recommends the following approach for selecting areas of influences that affect all the needed basic varieties of the field:

– Establish the type of studied random process. For example, the stationary process is determined by the dependence of the normalized correlation using only the difference in variables.
– Establish the basic characteristics of this process. For a stationary random process, we must have the mean, standard deviation, normalized correlation, and power spectrum.

– Define an area's ergodic, the possibility to make judgments about the process from one realization. This is when the correlation approaches zero if the time $\tau \to \infty$.
– Check the hypothesis that the process is normal. Try the Pearson or other criteria.
– Calculate the length of the influence area.
– Select the size of divergence between basic characteristics of different areas.
– Minimize the selected measure of divergence, and find the area of influence which address all possibilities of the field.

If one cannot simulate the simultaneous and complete combination of the field input influences, one cannot obtain usable results after testing that provide accurate initial information for solving quality, reliability, fatigue, durability, and other problems.

The number and types of field input influences that must be simulated in the laboratory depend on the result of the analysis of the action of the field influences on the product degradation (failure) mechanism, which shows how the influences are linked with the product degradation.

Step 6: Selecting a representative input region for useful accelerated reliability testing

The content of this step can be seen in Sect. 4.4.

Step 7: Procedure of accelerated reliability/durability testing

The procedure should consider the following:

• The schedule of the product's technological process including all areas and conditions of use (time of work, storage, maintenance, etc.)
• Sensor capabilities and availabilities
• What kind of input influences and output variables occur under each field condition to be analyzed
• The measurement regimes of speed, productivity, output rate, etc.
• The schedule for checking the testing environment
• •Typical measurements of conditions and environments in the field, and their comparative simulation results by accelerated reliability/durability testing
• Establishment of the value of testing
• Execution of the testing
• Obtaining test results and analyzing the data

Step 8: Use of statistical criteria for comparison of accelerated reliability/durability testing results and field results

As was shown above, the degradation mechanism is a common criterion of comparing field results and accelerated testing results. To perform this comparison, one must use apply statistic that compares the desired reliability/durability requirement with the test results. For this goal we will use the term "statistical criteria."

Use of these criteria can help one to decide whether to use one's current ART/ADT technique or whether it is necessary to develop this technique until the measure of the difference reliability or other function distribution in ART and in the field is no more than the fixed limit.

The statistical criteria that are shown below must be used during the following two stages of testing:

- During ART/ADT for a comparison of the output variables or lab degradation process with degradation process in the field.
- After finishing ART/ADT for comparison of the reliability indexes (time to failure, failure intensity, etc.), as well as other parameters with field reliability indexes. Additional information might be generated as this step by analysis of failed and not failed test subjects.

The difference between the field and ART/ADT testing results for both situations should be no more than a fixed limit. The limit for fixed parameter differences in the field and in the ART/ADT depends on the level of desired accuracy such as 3%, 5%, or 10% maximum difference.

One criterion was shown above. One can use also one or any of the following basic criteria for correlating results of ART/ADT with field testing:

$$\left[(C/N)_L - (C/N)_f \right] \le \Delta_1$$

$$\left[(D/N)_L - (D/N)_f \right] \le \Delta_2$$

$$\left[(V/N)_L - (V/N)_f \right] \le \Delta_3$$

$$.....................$$

$$[(F/N)_L - (F/N)_f \le \Delta_i$$

where

$\Delta_1, \Delta_2, \Delta_3 \ldots \Delta_i$ are divergences calculated from the results of the laboratory and field testing;

L and f are ART/ADT and field conditions;

$C, D, V,$ and F are measures of output parameters corrosion I;

D is destruction of polymers, rubbers, wood, etc.;

V is vibration or tension; etc., failures; and

N is the number of equivalent years (months, hours, cycles, etc.) of exposure in the field (f) or laboratory (L).

For example, Fig. 4.9 shows normalized correlation data for $\rho(\tau)$ and power spectrum $S(w)$ of a car's trailer frame tension data from the field and from simulation after using the above criteria.

The author describes in [1] and [2] the implementation of criteria for correlating results of machinery ART/ADT with the field results. The results of the implementation of these criteria for a tension car's trailer field axles are shown in Tables 4.1 and 4.2.

Fig. 4.9 Normalized
correlation ρ(τ) and power
spectrum S(w) of the car's
trailer frame tension data:
_____ field; ------------
ART/ADT

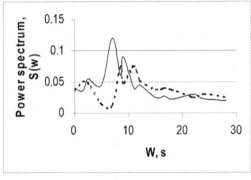

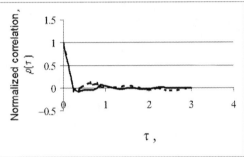

Table 4.1 Results of the calculation of tensions on the wheel axles of the fertilizer applicator

Class number	Class frequency N_i	Class accumulated frequency		Accumulated relative frequency			Modulus of congruence accumulated relative frequencies difference Max |D|
		N_j	$\sum N_i$	$\sum N_j$	$P_i = \dfrac{\sum N_i}{N}$;	$P_j = \dfrac{\sum N_j}{N}$	
1	10	16	10	16	0.020	0.040	0.020
2	18	10	28	26	0.070	0.065	0.05
3	20	14	48	40	0.121	0.100	0.021
4	32	32	80	72	0.202	0.180	0.022
5	39	50	119	122	0.303	0.305	0.002
6	84	68	203	190	0.512	0.476	0.037
7	68	72	271	262	0.680	0.655	0.025
8	49	63	320	325	0.808	0.812	0.004
9	30	36	350	361	0.884	0.900	0.016
10	24	16	374	377	0.994	0.940	0.004
11	10	12	394	389	0.969	0.971	0.002
12	12	11	396	400	1.000	1.000	0.000

Table 4.2 Part of assembly of fertilizer applicator's tension data

Digit number	Field N_i	Laboratory testing n_j	Difference $n_i - n_j$	Definition from average $[(n_i - n_j) x]$	Definition from average squared $[(n_i - n_j) x]^2$
1	1	1	0	−32.5	1056.25
2	5	4	−1	−33.5	1122.25
3	18	8	−10	−42.5	1808.25
4	20	25	−5	−27.5	756.25
5	121	66	−55	−87.5	7658.25
6	223	174	−49	−81.5	6642.25
7	270	201	−59	−101.5	10302.25
8	217	107	−110	−142.5	20302.25
9	105	40	−55	−97.5	9506.25
10	37	11	−26	−58.5	3422.25
11	13	4	−9	−41.5	1722.25
12	3	1	−2	−34.5	1190.25

$\sum_{ni} = 1033$ $\sum_{nj} = 642$ $\sum_{ni - nj} = 391$ $\sum_{[(ni - nj) - x]}^2 = 65489.5$

Figure 4.10 demonstrated the graph of normalized correlation for the shaft of the bottom spreading unit loading processes in the field (F) and the different regimen of testing in the laboratory.

Let us show a simple example of how we can find this correlation. Figure 4.11 shows experimental distribution of tension frequencies on the fertilizer applicator axle (sensor 1), the frame (sensor 2), and carrier system (tension 3) during fertilizer distribution in field conditions and in laboratory testing.

The results of calculation based on sensor 1 are in Table 4.1. These comparison data show very little deviation.

Here, i is in the field and j is in the ART/ADT:

$$\lambda = \max |D| \cdot \sqrt{\frac{\Sigma N_i \cdot \Sigma N_j}{\Sigma I + N_j}}$$

If λ is less than 1.36, we can accept the hypothesis about both samples being linked to one statistical population, i.e., the loading regimes in the field and in the laboratory are closely related to each other. Here $\lambda = 1.36$ is the tabled value of Smirnov's criterion at the 5% level.

Table 4.2 shows an example of using Student's distribution for evaluation of the mean. These results were obtained by sensor 2 in the field and in the ART/ADT (laboratory). In this example, the definition of average is equal to 1.4. Therefore, 1.4 < 1.8, where 1.8 is significant from the tables of Student's criterion at the 5% level (this information can be found in books on additional the theory of probability).

Therefore, our hypothesis is true for this example also.

The results of comparison random tension data in the field and in the ART/ADT conditions are illustrated in Fig. 4.12, which shows normalized correlation

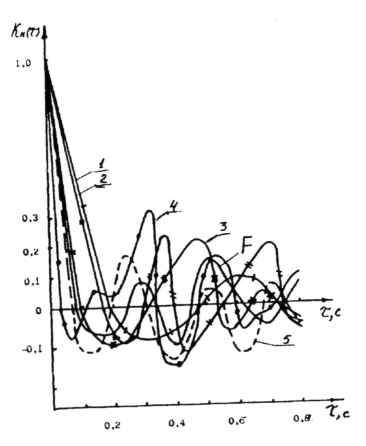

Fig. 4.10 Normalized correlation functions of shaft of milling unit loading processes (F, field; 1, 2, 3, 4, and 5 are numbers of the regimen during the laboratory testing)

functions and power spectrums of the car trailer's frame tension data requested by sensor 1. The time of correlation is between 0.1 and 0.12, the time of attenuates is the same, the power spectrum maximum corresponds to frequencies of 8–12 s, and the interval of frequencies is substantially the same (from 0 to 16–18 s). The power spectrum maxima have low velocity. Overall the regime of testing the carrier and running gear systems of the fertilizer applicator in the laboratory is closely related to the test regimes in the field.

Step 9. Collection, calculation, and statistical analysis of accelerated reliability/durability testing data

This step concerns test data collection during the test time, statistical analysis of this data based on the failure (degradation) type and test regimes, and the reasons for deterioration leading to the ultimate failure of the test subject, counting the accelerated coefficient, etc.

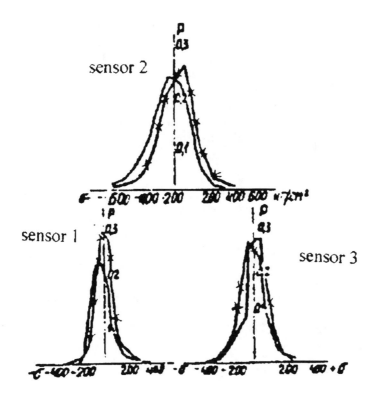

Fig. 4.11 Distribution of fertilizer applicator tension amplitudes in the field and in the ART/ADT conditions (-x- is the field, – is the ART/ADT conditions)

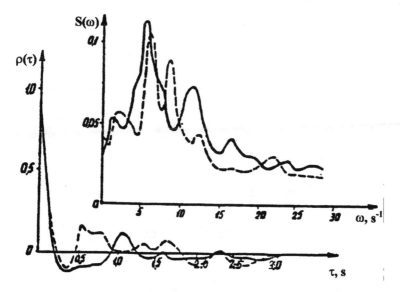

Fig. 4.12 Normalized correlations $\rho(\tau)$ and power spectrum $S(\omega)$ of car trailer's frame tension data: _____ field; - - - - - - - - ART/ADT

Figure 4.13 shows an example of accelerated loss of paint protection in the test chamber. One can easily find the acceleration coefficient, as is shown below.

The loss of protection quality of paint A (curve 1) for 5 days of ART/ADT is the same as for 90 days in the field, and the accelerated coefficient is 18.

Since the paint deterioration data show the importance of collecting and analyzing many data points, most ART/ADT are monitored by computer and data automatically collected, as this can be easily set up. A second important function of the computer is to ensure the test conditions are maintained, or to monitor deviations from the desired conditions. The importance of this data collection step is discussed by Chan and Parker [26].

Step 10: Prediction of the dynamics of the test subject's efficiency (reliability, durability, maintainability, and other components) during its service life

ART/ADT is not a goal; it is a source of obtaining initial information for efficiency (quality, reliability, maintainability, durability, life cycle cost, and other components), successful prediction, and problems solving. One can read in Chap. 3 how one can do it.

If the ART/ADT results have sufficient correlation with the field results, it is possible to rapidly find the reasons for test subject degradation and failure.

These reasons can be found by analyzing test subject degradation during time of usage, through the location of initial degradation and the development process of

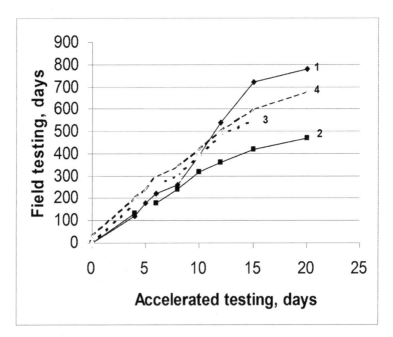

Fig. 4.13 Accelerated destruction of paint protection in test chambers during accelerated stress testing (two types of paint): first type of protection (paint A), 1 = protection quality; 2 = impact strength; 3 = bending strength; second type of protection (paint B), 4, impact strength

degradation. Then, it is necessary rapidly to eliminate these reasons. This method is time and cost-effective. Figure 4.14 shows, as an example, the process of accelerated reliability improvement.

If the above correlation is not significantly close enough, the reasons for the test subject degradation and the character of this degradation during ART/ADT usually do not correspond to this degradation that are found in the field. Therefore, the conclusion for ART/ADT results may be wrong and increase the cost and time of test subject improvement and development.

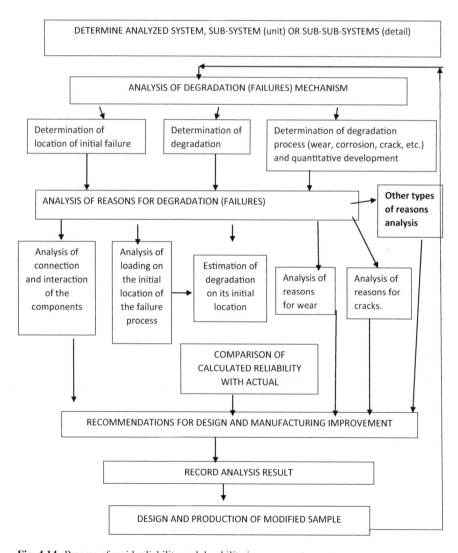

Fig. 4.14 Process of rapid reliability and durability improvement

In this case, the designers and test engineers often think that they have understood the reason for failure (degradation), and change the design or manufacturing process. But after checking the "improvement" in the field, they will know how wrong they are. Then, they should seek other reasons for degradation, etc. This process doesn't allow rapid improvement or development of test subject reliability (quality, etc.). Moreover, it only increases the cost and time of test subject development and improvement. This situation is a familiar one in practice.

One can read in Chap. 6 the two examples illustrating the implementation of the authors' approach to ART/ADT for rapid improvement of different product quality/ reliability/durability.

This concerned a problem with a harvester's reliability, which designers failed to eliminate from field data testing over several years. A laboratory complex was built that accurately simulated the field input influences on the harvester.

If we do not have accurate initial information which generating the results of ART/ADT, even the best methods of reliability prediction cannot be useful.

4.4   Basic Aspects of Methodology for Selecting Representative Input Regions for Accurate Simulation of Real-World Conditions

4.4.1   The Problem

As was mentioned in Chap. 3, there were no methodologies for selecting a real-world representative region that could be used for simulation in the laboratory until the author solved this problem. The full methodology is published in [2]. The basic aspects of this methodology are dealt with in the following discussion.

The real-world conditions for most products are used in a wide variety of environmental climatic zones. The selected regions (zones) will feature different environment, mechanical, and other influences, road surfaces, loads, pollutions, or workloads. As a result, the efficiency components in different regions will be vastly different. Therefore, we need to establish representative regions of input processes that correspond closely to the field influences, such as air temperature, humidity, pollution, radiation, and so on, based on the estimates of the output processes, such as measures of the output parameters of degradation, including tension, vibration, corrosion, and so on. The representative region approach is based on the evaluation of statistical characteristics of input (or output) processes, which thus presents the possibility of determining one region with sufficient characteristics and statistical information about the multitude of regions that exist and simulating its characteristics in the laboratory. The methodology has been motivated by and applied successfully to the real accelerated tests and is based on the theory of random processes. It is clear that this work must be carefully performed to have the desired results. The basic aspects of the methodology are given in the next section and are followed by an illustration of its application leading to a successful prediction.

4.4.2 Basic Steps of Methodology for Selecting a Representative Region

The methodology is summarized in the following steps.

Step 1: *Algorithm*

The algorithm for selecting a representative region for the accelerated test consists of the following substeps:

Substep 1: Identify the type of process exhibiting a random rate, and establish whether it is a stationary process through the:

- Characterization of the process in terms of its mathematical expectation, standard deviation, correlation, and power spectrum.
- Evaluate whether the process is ergodic by assessing whether the important correlation tends to zero as time tends to infinity.
- Use the Pearson criterion, or equivalent, to assessing whether the process exhibits normality.

Substep 2: Measure the divergence between the basic characteristics of the conditions evaluated in different regions.

Substep 3: Select and measure a characteristic in a representative region (usually the length).

Substep 4: Identify the representative region that has minimal divergence.

Step 2: Selection of a Process and Its Characterization

We assume that there is an interval $[0, T]$ in the length of written signal from the sensor (oscillogram or others) that characterizes the changes in loading or wear (or another physical parameter, X) of a product during use. We then divide the interval $[0, T]$ of the oscillogram into smaller intervals of length T_p, and for each, we evaluate mathematical expectation (μ_X), correlation function ($\rho_X(\tau)$), power spectrum ($S_X(\omega)$), and their accuracy using, for example, Lourie method [28]. Thus, the periodicity of the process can be checked to ensure that it is appropriate to use one interval realization of the process as being representative of all realizations of the same duration. Periodicity can be assessed formally by checking that the correlation function tends to zero as $\tau \to 0$; it can be used to analyze the ergodicity of the process. However, it is also appropriate to combine this information with judgment relating to particular problems and the physical essence of the process being studied.

To check the hypothesis about the normal character of a process, we assume that using one realization of the specified interval length T_p is sufficient to characterize the entire process because it is ergodic. From the correlation function, we can identify the value of τ for which the correlation is (statistically) zero, that is, ($\rho_x(\tau_0)$). For the interval $[0, \tau_0]$ in the oscillogram, we measure the values at equidistant points t_1, t_2, $t_3 \ldots t_n$. The usual goodness-of-fit tests, such as Pearson, can be used to assess whether these values are normally distributed.

Step 3: Length of the Representative Region T_R

If it is established that the process is normal, we can begin evaluating the length of the representative region, T_R, which has to be large enough to evaluate the correlation function ($\rho_X(\tau)$) and the mathematical expectation because of the accuracy of μ_X. The interval representing T_R is divided into n equally spaced subintervals of length $\Delta t = T_R/n$, and the observed values of the oscillogram noted at different times are denoted by $x(t_i)$. Hence, the correlation function can be estimated using

$$\hat{\rho} = \frac{1}{m-n} \sum_{t_i=1}^{n-m} \left[x(t_i + m) - \mu_X \right]\left[x(t_i) - \mu_X \right] \tag{4.1}$$

To calculate the preliminary values of T_R, we need to know the frequency ranges in the oscillogram where the parameters are changing. Low-frequency periods, denoted by $T_1, T_2, T_3 \ldots T_n$ and each of which have a random rate, are marked at equal intervals from each other. The mean, denoted by T_M, can be evaluated by $T_M = {}^{T}1 \pm {}^{T}2 \cdots \pm {}^{T}n$ where $n \geq 3$.

The lowest frequency of process rate is given by $f_L = 1/T_M$, and the value of τ_{max} can be selected using the condition $\tau_{max} \approx 1/f_L = T_M$. The value of T_R can be estimated using τ_{max}. The typical value is 4, because the error associated with the estimation of μ_x and $\rho_x(\tau)$ should be no more than 5%.

To estimate Δt, it is usual to consider the highest frequency changes of the process, which we denote by f_M, and in practice $\Delta t = 1/0.6 f_m$.

Step 3: Comparison of Values Between Different Regions of the Oscillogram

Since stationary normal processes can be characterized by their expectation (μ), variance (σ^2), and the normalized correlation structure ($\rho_N(\tau)$), we evaluate these for the entire oscillogram over the range $[0, T]$ and for k different regions of the oscillogram of length T_R, where $k = T/T_R$.

The divergence between any two regions, say a_i and a_j, may be estimated as a weighted linear combination of the squared differences between their expectations, variance, and correlations. For example:

$$\Delta(a_i, a_j) = p_1 \left(\hat{m}_i - \hat{m}_j \right)^2 + p_2 \left(\hat{s}_i - \hat{s}_j \right)^2 + p_3 \max \left| \hat{r}_{Ni}(t) - \hat{r}_{Nj}(t) \right|^2 \tag{4.2}$$

where the weights, p_1, p_2, p_3 sum to one and are selected using the method of least squares to minimize the sum of squared deviations between the corresponding components of the two vectors. Hence, higher weights will be given the bigger differences between the corresponding statistics. If there is no difference between any of the components within the vector, then all will be equally weighted (i.e., 1/3). We also suggest using the maximum deviation between the correlations as a measure of the most extreme difference between the two correlations. This only uses a selection of information in the correlation structure. To overcome this, it may be preferred to use the estimated spectral power function $(S^\wedge(w^j))$ instead of the correlation function. In that case, the deviations between the two regions would be written as

$$\Delta\left(a_i,a_j\right) = p_1\left(\hat{m}_i - \hat{m}_j\right)^2 + p_2\left(\hat{s}_i - \hat{s}_j\right)^2 + p_3\left|\hat{S}_i\left(w\right) - \hat{S}_j\left(w\right)\right|^2 \qquad (4.3)$$

where again $p_1 + p_2 + p_3 = 1$.

Step 3: Choice of the Representative Region

The representative region is selected as the one, from the k considered regions, that has the minimum divergence from the whole range considered; that is, the representative region will be denoted by the vector *aR* and is the one that minimizes the corresponding function written first for the correlation structure and next for the power spectrum [29]. Thus, the region of length T_R that satisfies the aforementioned condition is selected to be simulated during accelerated testing, as this provides a representative selection of the oscillogram of reduced length [i.e., $(0, T_R)$ rather than $(0, T)$] and will facilitate more efficient testing.

For example, consider the results presented in Table 3.4 [2] for the mean, variance, and standard deviation, which show that there is a considerable spread in the estimates. However, the same is not true for the correlation. In contrast, the estimates across regions show that the maximum deviations in the normalized correlation are almost equal, which suggests that the process does reduce to stationary status.

The analysis of the power spectrum over the interval of ω from 0.9 to 2.5 s^{-1} covers most of the spectral density, whereas the correlation function for different regions, shown in Fig. 3.5 [2], appears to have a periodical component. Therefore, the correlation should take in a larger interval, such as from 0 to 10 s. Table 3.4 [2] shows the modulus of mathematical expectations, standard deviations, correlation, and spectral functions and the divergences from different regions based on the characteristics of the whole oscillogram. The divergences presented in the above table are substantial. An alternative way of examining the deviations between each region and the whole oscillogram could be to superimpose the corresponding functions on the same graph.

One can see in [2] the full above methodology with illustrations and examples.

The purpose of the above is to show an approach to the methodology of selecting a representative region from the multitude of input influences based on input (output) influence simulation. The result leads to successful ART/ADT. The concept of representative region selection is then extended to real-life analysis techniques to establish the characteristics of the critical input (or output) processes. This leads to the possibility of selecting one region from a multitude of regions, which is most characteristic of all statistical population of the field varieties.

This methodology helps to obtain accurate information from ART/ADT results for successful efficiency prediction.

4.5 The Role of Real-World Conditions' Accurate Simulation in the Development of Accelerated Reliability and Durability Testing and Successful Efficiency Prediction

The accurate simulation of real-world conditions is the basis for providing accelerated reliability/durability testing and successful efficiency prediction. As can be seen from Fig. 4.15, successful prediction of product efficiency depends on the level of development methodological aspects of prediction, as well as the level of development accelerated reliability and durability testing. Both development levels depend on how accurate is simulation of real-world conditions. The accuracy of this simulation influence of how methodological aspects of prediction reflects the real-world conditions. The same relates to accelerated reliability and durability testing, which level results depend on how accurate simulated the real-world influences, as well as human factors and safety problems.

The above aspects are considered in detail in author's book [1].

Another important aspect of real-world accurate simulation that can be seen is how the conduct of accelerated testing done during the development of finished products correlates with the development of new testing methods and equipment. Shown below is one example of the evolution of real-world simulation for testing together with the evolution of the test subject. In land, sea, and air applications, the

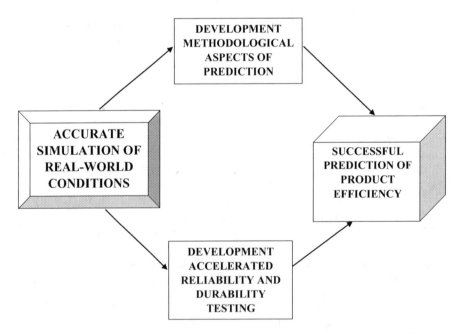

Fig. 4.15 Scheme of the role of accurate simulation of real-world conditions in the development of successful prediction of the product efficiency through

use of Unmanned Aircraft Systems (UASs) has shown and continues to demonstrate explosive growth with no sign of slowing. But the UAS faces a number of key challenges which must be addressed if it is to satisfy the future need of the American Department of Defense (DoD), other governmental departments, nongovernment organizations and companies, and individual consumers for UASs.

There are two paths that can lead to either decreasing or increasing the technical and economic characteristics in an engineering product or technology. Both of these paths begin with the accurateness of the simulation of real-world conditions for enhancing research, design, and manufacturing projects.

The benefits of accelerated reliability/durability testing are based on level of real-world simulation supporting this testing. This relates the connection of accelerated testing level with its benefits, related to recalls and other problems solution.

As can be seen in Fig. 4.16, accurate simulation and based on accelerated reliability and durability testing (ART/ADT) leads to dramatic decreases in recalls and can save multibillions of dollars. Following the opposite path of traditional accelerated life testing (ALT) can lead to losses of multibillions of dollars, as it was shown in Introduction. Moreover, if this trend and situation continues, it will engender more costly and more dangerous future results.

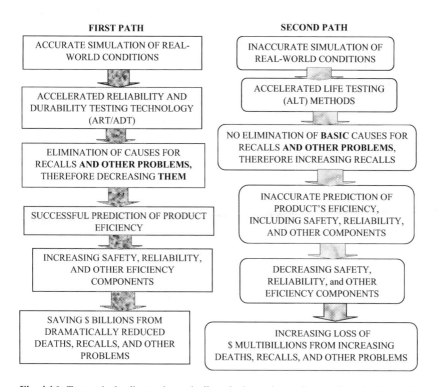

Fig. 4.16 Two paths leading to dramatically reducing or increasing recalls and other results

From this perspective, the importance of stopping the development of negative trends in accelerated testing can be understood. The strategy and methodology of doing this are detailed in [20].

These challenges to fulfilling the first path (way) are included in [30]. These challenges related to numerous areas of engineering.

For a better understanding of what is accurate simulation, accelerated, and especially reliability/durability testing, one must understand the effects of influences from real-world conditions, especially of product degradation. From the following figures, let us consider the types of degradations and their parameters as examples of this information requirement (Fig. 4.17).

The degradation parameters are interconnected in real life. The example of mechanical degradation parameters is demonstrated on Fig. 4.18.

The development of accelerated testing in engineering (and not just in engineering, but in many other areas of technology) requires the development of new testing technology (methods and equipment) used for obtaining accurate initial information that enables successful prediction of product efficiency. This technology must consider the content of the above figures. The ART/ADT correspond the above aspects; therefore, it is appropriate.

If we would consider the role of accelerated testing as a source of interactive initial information about product efficiency, we could not only successfully improve product testing, but we would also improve product efficiency and organizational

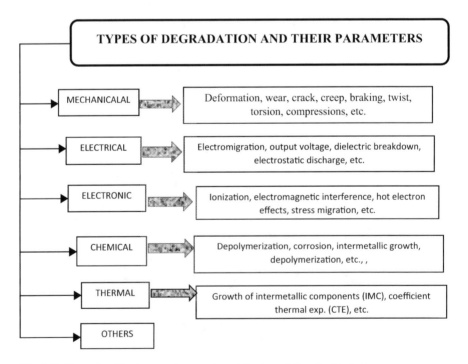

Fig. 4.17 Example of the types and parameters of the degradation mechanisms

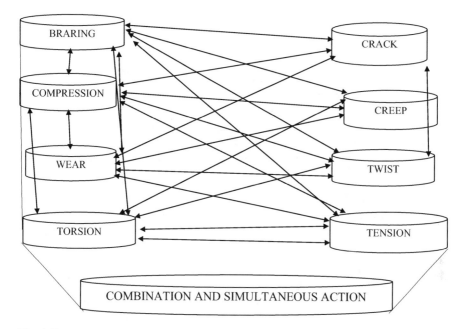

Fig. 4.18 Interconnection of different types of mechanical degradation

economic efficiency. Too often the traditional approaches to new product development have considered accelerated testing (and testing in general) as a distinctly separate discipline. Thus, it is done without connection to the other components of the product's creation, design, and manufacturing efficiency. While this is not to suggest that accelerated testing is a discrete discipline, this development does need to take into account the cited elements (Figs. 4.6 and 4.7).

Finally, it must be membered that many of the statistical approaches to accelerated testing are primarily theoretical aspects of estimation, which do not consider the real nonstationary random character of the processes that are experienced in real-life operation.

Therefore, ART/ADT will be more beneficial for producers and users alike, because it eliminates most of the negative aspects of what is now termed accelerated testing.

The Role of Engineering Simulation

Based on the historic trends in design and testing that have been observed and the Unmanned Aircraft System (UAS) [31] roadmap laid out by the major users, several key design constraints in the development of future UAS platforms and payloads can be expected, to include:

- Very short development cycles
- Short-term design customization with little design precedent
- Medium- to long-term design optimization for standardization

- Increasingly complex missions with associated capability innovation and integration

Engineering simulation harnesses the power of computers and software to solve the fundamental equations of physics or those that are close approximations of these equations. This allows designers and test engineers, and others involved in analysis to create virtual representations of complete UASs and their payloads for design analysis and optimization prior to physical testing. Correct implementation of the technology has been verified and validated in a range of industry sectors, and the use of more accurate engineering simulation is, in some cases, mandated by regulatory bodies.

Research results [31] have shown that best-in-class companies can:

- Meet quality targets 91% of the time, compared with a 79% industry average
- Meet cost targets 86% of the time, compared with a 76% industry average
- Launch on time 86% of the time, compared with a 69% industry average

The outstanding difference in strategies pursued by first-class companies is the systematic use of engineering simulation (physical and software) regularly throughout the design and testing process. In essence, these companies consistently leverage engineering simulation throughout the design process, with accelerated testing as one of the basic components of this approach. Those lead to improvements in quality, reliability, cost, time, and other components of performance when compared to companies that do not do this.

Research performed by the US DoD revealed the staggering impact that engineering simulation can have [32] and [33].

A 3-year study reported that "…**for every dollar invested in accurate simulation the return on investment is between \$6.78 and \$12.92". There are recorded returns of between 678 percent and 1, 292 percent"** [33].

The above simulation will add the most value when:

- It is applied to all aspects of design (pre-design and design with accurate simulation of field conditions, accelerated reliability, and durability testing, not just one or two influences taken in isolation).
- The interaction of the physics at a system level is included in the analysis (e.g., analysis of action of the full real-world input influences).
- The workflow is seamlessly integrated across the physical essence and with existing tools.
- Physics-based optimization is performed across the design envelope.

At an organizational level, it needs to be recognized that engineering complex accurate simulation tools need to offer more than just technical capability. The unique nature of the completed products with their lack of design precedent makes it critical to capture the whole design process and intent. By doing this, it can be systemized and scaled for future applications. Capturing and managing this engineering knowledge is best performed by using the successful prediction methodology and tools, as described in this author's publications, for example [1], [2], and [20].

The ideal scenario for this is when a successful prediction methodology, including accurate physical simulation of the real-world conditions, accelerated reliability and durability testing (ART/ADT), and prediction methodology (all in one complex tool), performs and improves the engineering culture of management to provide one entire system producing only the right type and needed information.

This approach relates to an interdisciplinary systems of systems approach.

The close collaboration between OEMs and suppliers required for successful platform and payload integration demands the easy exchange of successful prediction while mitigating mutual intellectual property and data security concerns.

Having considered the growing complex needs of society and the benefits of accelerated testing technology as an interactive component of these needs, it is clear that advanced accelerated reliability and durability testing will be a fundamental enabler for the development of next-generation systems.

Finally, the role of accurate simulation can be seen in Fig. 4.19. As it demonstrates, inaccurate field/flight simulation leads to a poor level of testing. The result of this testing is inaccurate prediction of the product or technologies' effectiveness which then leads to low safety, reliability, durability, and other efficiency components.

This figure explains some basic reasons for the decreased economic benefits obtained from the new equipment or technologies, and why unplanned costs of recalls, crashes, people's deaths, and other negative outcomes can exist.

4.6 Establishing the Concepts and Statistical Criteria for Providing Physical Simulation of Input Influences on a Product for Accelerated Reliability/ Durability Testing

The author has established how a product's field conditions, including input influences, must be accurately simulated so us to provide the higher correlation necessary between the accelerated reliability or durability testing results and the actual field results. This is a fundamental requirement for obtaining successful prediction of real-world reliability, durability, safety, maintainability, and other components of efficiency prediction through the results of effective ART/ADT.

But exactly what are the needed information to provide successful efficiency prediction? The answer to this question can be found in this author's books [1] and [2]. In general, it has been his experience that the most accurate physical simulation of the field conditions processes occurs; then, each statistical characteristic of these processes—mean, standard deviation, power spectrum, and normalized correlation $[\mu, D \rho(\tau,) \text{ and } S(\omega)]$—differs from the real-world condition measurements by no more than 10%.

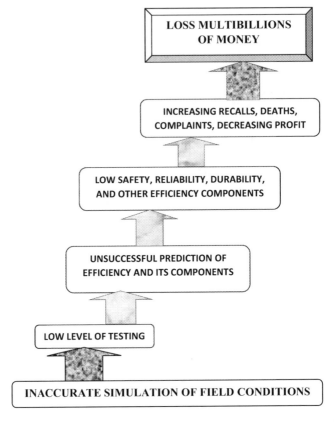

Fig. 4.19 The path from inaccurate simulation of field conditions to increasing recalls, deaths, complaints, decreasing profit, and other economic losses

One must calculate and use these statistical criteria for each specific situation to achieve this goal of obtaining acceptable correlation between these processes measured in an accelerated reliability/durability testing as compared to field processes. But all these analytic comparisons must include acceleration coefficients, because ART/ADT promise offers to obtain the results faster than in the real world.

The degradation process in the real world (field) and during ART/ADT will determine the practical limit of the applied testing stress, and acceleration coefficient. For establishing the accurateness of simulation, the author developed common criteria (see Chap. 3).

The first component of ART/ADT is met by increasing working time to the maximum extent possible. This type of acceleration can be used in testing if the product does not operate continuously (24/7), but can be tested 24 h a day, every day. Generally, it does not include idle time or operating time with minimum loading. This method is based on the principle of reproducing the complete range of operating conditions and maintaining the proportion between heavy and light loads. The author's experience shows that this method of acceleration has the following basic advantages:

(a) Good correlation between field and laboratory testing results.
(b) Each hour of clear of productive work performed by the product is faithfully reproduced in the stress schedule which is identical to the destructive effect of 1 h of productive work under normal operating conditions.
(c) There is no need to increase further the magnitude, size, or pace of the applied stress in the testing than what has already been described.

This accelerated testing generates good data faster than in the field by ten and more times. This method is especially useful for a product which works for short periods of a time (during the day, month, year).

Professionals who use this method know that reproducing a complete range of operating conditions is not always easy, but it yields a more successful correlation of the accelerated testing results and the actual field results.

Test conditions are designed to include a full array of different parameters (temperature, humidity, vibration, radiation, and others) as experienced by the product in the field.

When it is necessary to obtain faster and simpler accelerated testing results, the increased stress methods of accelerated testing can be employed. However, when utilizing this method, the correlation between the laboratory test results and the field results will be lowered. Higher stress means less correlation, which means less accuracy of simulation and greater problems with accurate evaluation and prediction of product reliability, durability, and other efficiency components in the field.

With lowered correlation there also are more problems in finding the true reasons for failures or product degradation, and the correct solutions for eliminating them. Thus, for any specific product, one must decide which stresses to evaluate and how much increased stresses for products will result in be more effective tests.

Acceptance of how fully the accuracy of any method of simulating real-life input influences and their simultaneous combinations depends on the testing goal. If it is necessary to have an independent simulation, for example, for providing vibration testing without other influencing factors, the results of this testing lead to not accurately predict the reliability or durability, or other components of the product. The same is true for simulation of temperature (or temperature/humidity) testing. Temperature (or temperature/humidity) testing is not truly environmental testing, because temperature and humidity are only two parameters out of the plethora of many possible field environmental influences that are experienced by the product during normal usage.

4.7   Equipment for Advanced Accelerated Testing

4.7.1   Introduction

As was discussed earlier, usage of virtual approaches to accelerated testing is increasing, and the physical approaches to accelerated testing are decreasing. As the basic technology of testing consists of methods and equipment, the methods and speed of physical testing development are decreasing. Likewise, the speed of the equipment development, needed for physical accelerated testing, is also decreasing. Often, the modern equipment being developed for accelerated testing has only been concentrated on new control that is based on electronic technology development. To see this, one must do a comparison of Test Expos for the automotive and other industries now and of previous years.

The same companies that tried to develop equipment for more complicated equipment for accurate physical simulation of field conditions can now often be seen moving back to simpler equipment and less accurate simulation of real-life conditions.

If you look at the accelerated reliability and durability testing equipment produced by companies, such as Weiss Technik, Instron Schenck, Advanced Test Equipment Corporation, WestTest, LDS, RENK, Horiba, and others, you will see little in the way of new equipment for physical testing of entire machinery, but a lot of equipment for materials and details testing without interconnection factors, or programming for virtual testing.

And while you will frequently see the term "durability testing" or "reliability testing," in fact, most of this is proving ground testing or fatigue testing, or some other types of testing, but it is not accurate to call it as "durability testing" or "reliability testing."

The reason for this situation is not with the companies that design and manufacturing testing equipment, but with their clients whose interest too often is primarily in saving money in the testing area. Because virtual testing is less expensive, many companies in different industries want to transition to it. While they believe that simple and less expensive testing will still lead to accurate simulations of field conditions, they fail to consider increased expenses of increased maintenance and product recalls after the design and manufacturing life stages. Too often the final result of simpler testing is decreased quality, reliability, durability, maintainability, profit, and increasing life cycle cost.

As can be seen in [20], while there is an overall trend of an increased market for testing equipment, closer analysis this increase relates it mostly to:

- Increasing development in the electronics industry
- Increasing development of apparatus and instruments
- Increasing development of the materials and details of the product

But what about the market for equipment for testing entire cars, trucks, buses, airplanes, space research stations, and other machinery? What is being done for the other aspects of engineering technologies testing development? What is being done to develop equipment for accelerated testing of units and complete product such as cars, trucks, satellites, and many others? The author presents a partial analysis of this in Chap. 4 "Equipment for Accelerated Reliability Durability Testing Technology" (74 pages in the book [2]), and his paper 2022-01-012 for SAE 2022 World Congress. Essentially, it shows how most changes have been focused on system of control testing equipment, and such systems primarily connected with development in electronics.

What is being done for the other aspects of engineering technologies testing development? What is being done to develop equipment for accelerated testing of units and complete product such as cars, trucks, satellites, and many others?

There is a real need for more effective and successful testing equipment in these new technologies, especially successful prediction efficiency in these new technologies. Otherwise, how does one obtain the initial information needed for successful prediction the efficiency of a new product and its components, including quality, reliability, safety, durability, maintainability, life cycle cost, and many others? This author commented these problems in detail in his books [1–3, 20], articles, and papers.

Unfortunately, the managers of many industrial companies do not appreciate the simple fact of life that new product development for more complicated products needs correspondingly more complicated testing equipment. Solving this failure requires real leadership that invests money into proper equipment that will lead to long-term better products and higher profits.

4.7.2 General Situation

This section of the chapter will summarize the author's findings about the general situation in the development of testing equipment in the world market.

Technavio's latest report on the global general-purpose test equipment market provides an analysis of the most important trends expected to impact the market outlook from 2017 to 2021.Technavio's latest report on the global general purpose test equipment market provides an analysis of the most important trends expected to affect the market outlook from 2017 to 2021. Technavio [34–36] is a leading global technology research and advisory company. The company develops over 2000 pieces of research every year, covering more than 500 technologies across 80 countries. Technavio has about 300 analysts globally who specialize in customized consulting and business research assignments across the latest leading edge technologies.

These analysts employ primary as well as secondary research techniques to ascertain the size and vendor landscape in a range of markets. They obtain information using a combination of bottom-up and top-down approaches, as well as using in-house market modeling tools and proprietary databases.

They corroborate these data with others obtained from various market participants and stakeholders across the value chain, including vendors, service providers, distributors, resellers, and end users.

General-purpose test equipment (GPTE) includes different testing and measuring (T&M) equipment such as oscilloscopes, spectrum analyzers, signal generators, power meters, logic analyzers, electronic counters, and multimeters. Technavio forecasts the global general-purpose test equipment market to grow to 6.58 billion USD by 2021, at a compound annual growth rate (CAGR) of nearly 5% over the forecasted period.

Below is reviewed the automotive and aerospace testing equipment that relate to other industries as well.

The Automotive Test Equipment Market [35, 36]
The automotive industry until now have not had widely used accelerated reliability and durability testing methodologies. They have used mostly separate components for this testing or combined testing, where one simulates just a few of many significant real-world influences. Therefore, they mostly cannot make accurate prediction of product efficiency.

One can read in detail overview of this testing equipment in [2]. The automobile test equipment industry often presents complexities not found elsewhere. New designs of automotive products, shorter vehicle development time to market, and increasingly stringent emission control regulations by the government are all creating new challenges for the automotive industry. In order to address such challenges, automotive manufacturers must acquire new and more sophisticated automotive test equipment, thereby driving the growth of the T&M market.

The equipment testing market is gaining traction in the areas of rheology fire mechanics, media resistance, and surface performance. Automobiles are increasingly being equipped with modern systems and complex electronic safety devices and systems, which are generating greater needs for the adoption of automotive test equipment. The percentages of electronic devices in automotive products is increasing at a very fast pace with many mechanical components being replaced by electronic components.

This generates a corresponding increase in the demand for suitable technologies to ensure that these new components are being adequately tested before being incorporated into automobiles. Robert Bosch GmbH (Germany), Honeywell International Inc. (United States), Siemens AG (Germany), ABB Ltd. (Switzerland), Delphi Automotive PLC (United Kingdom), Actia S.A. (France), Advantest Corp (Japan), Horiba Ltd. (Japan), Softing AG (Germany), ACTIA Group (France), EM TEST (Switzerland), Freese Enterprises Inc. (United States), Moog Inc. (United States), Sierra Instruments (United States), and Teradyne Inc. (United States) are some of the prominent players profiled in Market Future Research Analysis (MRFA) and are at the forefront of competition in the Global Automotive Test Equipment Market; included among these is Freese Enterprises Inc. a specialist in the automotive testing equipment market. They provide equipment for testing applications such as airbag controller, automotive paint and coatings, lighting systems, bumper stiffness, electric motors, electronic throttle, multifunction switches, and instrument panel cluster gauges.

The automotive test equipment market can be segmented into the following key dynamics for convenience of analysis and enhanced understanding of the market. These key segments are:

- Segmentation by product type, e.g., wheel alignment tester, engine dynamometer, chassis dynamometer, and vehicle emission test system
- Segmentation by vehicle type, e.g., light commercial vehicle, heavy commercial vehicle, and passenger cars

- Segmentation by application, e.g., mobile device based scan tool, handheld scan tool, PC-/laptop-based scan tool, and others
- Segmentation by regions, e.g., North America, Europe, APAC, and other specified regions

Charles Sweetser in [37] wrote:

The industry is always looking for better methods and techniques for confidently determining the condition of power transformers. Maintenance practices and philosophies are always being scrutinized and re-evaluated in hopes of maximizing diagnostic value and balancing economic efficiency. Traditionally, our industry has practiced conventional off-line tests which depend on a single measurement at a single frequency, constant voltage, or constant current.

Often, only having conventional test data for review has resulted in inconclusive analysis that often leads to more unanswered questions. The industry is demanding reliable diagnostic information that is representative of the best possible condition estimation. In 2013, the emerging trend is to extract as much additional diagnostic information as possible by applying smarter advanced methods and techniques to existing procedures. This will require using multi-functional test instruments with advanced features. The idea is not to create new tests and increase testing overhead. Varying parameters to conventional tests, such as frequency, provide a new avenue for analysis. Based on research, practical experience, and advancement in measurement instruments, it is now possible to extract in-depth information that was not available in the past. These advanced diagnostic methods or "extensions" provide new and critical information about the transformer condition.

Mark Schrepferman wrote [38] that:

… Test & Measurement system designers will be challenged to provide faster, extremely repeatable, rugged, and best-in-class testing environments. Test equipment will be expected to last over multiple generations of product introductions, meaning that the performance requirements of the RFICs used in test equipment must be better than the device under test by a factor of several generations. Further to this, next-generation communication systems that use higher-order modulation schemes such as Orthogonal Frequency-Division Multiplexing (OFDM), with high peak to average ratios, are driving the need for the components used in the test equipment's signal chain to have higher linearity. Additionally, more frequency bands will be introduced, driving the need for broader bandwidths and higher operating frequencies.

Finally, despite the fact that test solutions are growing in complexity, end customers will continue to expect lower overall test costs per unit.

And, as Mike Fox wrote in [39]:

A fresh new way of thinking about test and measurement is emerging. As we look to 2013, we are transforming testing and diagnostics with technologies that accelerate and improve the efficiency of diagnostic/repair/approval workflows and processes into what is a true "diagnostic ecosystem." In addition to the use of touchscreens with intuitive menu controls that emulate today's personal electronics, several of thermal imagers use Wi-Fi technology and mobile apps to connect to Android or Apple iOS tablets and smartphones. Electrical readings can be stamped right on the image. And wireless data streaming DMMs can share readings with PCs. Look for smartphone/tablet connectivity as well. The goal of the diagnostic ecosystem is not only to improve communication among diagnostic and communication devices but also among technicians and their customers and managers, by leveraging accurate and coordinated readings from related tools as well as rapid and actionable communication.

4.7.3   *Environmental Combined Testing Equipment*

Let us consider aerospace industry. This industry has historically not used accelerated reliability and durability testing. However, one can read below what testing equipment is currently being used by them.

Flight Hardware

When building spacecraft and associated space flight hardware, the design process most often requires a comprehensive set of testing capabilities. The testing's objective is to verify the ability of the design to survive its design loads, and to qualify the design for flight. Such tests frequently include static loads, model, vibration, and acoustic and shock tests. But one effect of this testing is the resulting extension of the time and increased cost required for putting the hardware aloft.

The idea of a combined testing has been developed over recent years with some real success, but it has received little attention in the technical media.

One such example is the Burke E. Porter Machinery Company (BEP) [40] which has worked in combination with vehicle test systems. As the manufacturer of the first fully electric dynamic vehicle testing (DVT) roll/brake machine, BEP has paved the way for innovation in automotive and heavy-duty vehicle testing, supplying end-of-line equipment, dynamometers, and tire and wheel assembly testing systems to all major OEMs throughout the world.

Environmental Testing and Test Facilities

European Test Services (ETS) [41] is maintaining and providing test facility services to European industries by managing and operating the environmental test center of the European Space Agency (ESA) located in Noordwijk, The Netherlands. ETS provides mechanical testing, electromagnetic compatibility (EMC) testing, thermal vacuum testing, altitude simulation, and more.

In addition to the testing of spacecraft and space applications, ETS has also become a major supplier of testing services to railway, marine, and the electric power industry [41, 42].

ETS Mechanical Data Handling Facilities

ETS's mechanical data handling facilities provide data acquisition, reduction, and presentation for vibration, acoustic, and shock testing. Their acquisition system can handle various analogue inputs and consists of four mobile/modular 128 channel racks interconnected to the test article via mobile patch panels.

In addition to this acquisition hardware, ETS can also provide acceleration sensors and a Force Measurement Device (FMD) to their customers. This FMD consists of load cells connected via a summation unit to the data acquisition system. In this way the interface forces between shaker and test devices are available during the test for control/notching and immediately available after the test for response analysis and model identification.

ESCO Technologies Company, ETS-Lindgren (subsidiary company), and Chamber, Enclosure, and Test Cell Solutions for Test and Measurement Applications ETS-Lindgren [43, 44] are working on systems and components for the detection, measurement and management of electromagnetic, magnetic, and acoustic energy. The company adapt technologies and applies proven engineering principles to create value-added solutions.

This company began in 1995 when EMCO, Rantec, and Ray Proof combined their resources to create a new entity called EMC Test Systems. Later, after this acquisition, the company changed its name to ETS-Lindgren.

In 2000, Holiday Industries was also acquired. The company's testing technology is used in today's electromagnetic field sensing (EMF) systems for test and measurement, and health and safety applications.

Continuing their growth in 2002, ETS-Lindgren purchased Acoustic Systems, a supplier of acoustic test and measurement, audiology, and broadcast applications.

Chambers, Enclosures, and Test Cell Product Gallery
EMC Chambers

Free-Space Anechoic Chamber Test (**FACT**)

Free-space anechoic chamber test (FACT) represents state-of-the-art technology for EMC measurement using demountable modular panels, anechoic absorber, and sliding, swing, or hinged doors. FACT chambers provide the test environment for meeting most international emissions and susceptibility standards such as CISPR, IEC, VCCI, ANSI, FCC, and SAE.

Statistical Mode Averaging Reverberation Chambers (**SMART**)

Statistical Mode Averaging Reverberation Test (SMART) chambers provide an electromagnetic environment for performing both radiated susceptibility and emissions testing. IEC 61000-4-21 draft standard addresses the reverberation chamber as a test environment.

SMART 80 provides these capabilities:

- 80 MHz–18 GHz frequency range
- Continuous or stepped tuner rotation
- Typical size for SMART 80 chamber: 13.44 × 6.09 × 4.87 m (interior)

Weiss Technik Chambers
Regarding electrical vehicle (EV) fast development, there are new trends under development, including work on EV/batteries. From many companies in testing equipment, one of the better ones is Weiss Technik. This company developed environment test chamber solution EV/battery [48].

1. *Walk-In and Drive-In Chambers*

If the standard test chambers are not large enough for the customers or the test requirements, there is a special solution, Weiss Technik offers almost unlimited options. As a single-source supplier, the company develops and implements test chambers and test rooms for modules, packs, and complete drive units, with or without BMS. In terms of size, there are choices ranging from walk-in test chambers up to test rooms for entire vehicles.

2. *Reach-In Series*

Special testing tasks require special test chambers. This is why it was modified the standard chambers according to the hazard assessment and requirement at hand, for example, by adding safety components such as a flushing device with a particularly high air replacement rate. In addition, it was offered a wide range of special solutions, such as positioning the control technology above the test chamber, for heavy-duty gratings with a telescopic system and drawer systems for up to 12 batteries with a guide-through and plug-in connector panel.

3. *Testing Safety*

Laboratory Hazards

Testing lithium-ion packs, modules, and cells with their increasing energy densities is a sensitive topic (Table 4.3). During the temperature tests, overcharging or malfunctions of the batteries may occur. This can lead to the destruction of the batteries. Increasing storage sizes cause increasing impacts of possible failures and potential risks during tests with lithium-ion batteries. For this reason, safety in the laboratory, in particular the protection of the staff during such tests, has the highest priority.

Tests Under the Influence of Temperature

External influences, such as

- External heating
- Overcharging

Table 4.3 Impacts of the lithium-ion battery

Hazard level	Description	Classification criteria and effect
0	No effect	No effect. No loss of functionality
1	Passive protection activated	No defect; no leakage; no venting, fire, or flame; no rupture; no explosion; no exothermic reaction or thermal runaway Cell reversibly damaged. Repair of protection device needed
2	Detect/damage	No leakage; no venting, fire, or flame; no rupture; no explosion; no exothermic reaction or thermal runaway. Cell irreversibly damaged. Repair needed
3	Leakage Δ mass <50%	No venting, fire, or flame*; no rupture; no exposition. Weight loss <50% of electrolyte weight (electrolyte = solvent + salt)
4	Venting Δ mass ≥50%	No fair or flame*, no rupture; no explosion. Weight loss ≥ 50 % of electrolyte weight (electrolyte = solvent + salt)
5	Fire or flame	No rupture; no exposition (i.e., no flying parts)
6	Rupture	No exposition, but flying parts of the active mass
7	Explosion	Explosion (i.e., disintegration of the cell)

- Deep discharge
- Excessive charging current
- External short circuit

Internal events, such as

- Electrode electrolyte reactions
- Electrochemical reactions

- The presence of flame requires the presence of an ignition source in combination with fuel and oxidizer in concentrations that will support combustion. A fire or flame will not be observed if any of these elements are absent. For this reason, it was recommended that a spark source be used during tests that are likely to result in venting of cell(s). Credible abuse environments would likely include a spark source. Thus, if a spark source was added to the test configuration and that gas or liquid expelled from the cell was flammable, the test sample would quickly progress from Hazard level 3 or 4 to Hazard level 5.

Weiss Technik's battery testing equipment and battery testing systems play a pivotal role in the development and production of computer equipment, smartphones, and electric vehicles. By utilizing Weiss Technik's battery testing equipment and battery testing systems, one can count on a productive life. But one cannot successfully predict the durability of test subject during service life, because this equipment is simulated only several from many real-world environmental components.

But this is not Weiss Technik's mistake; this is result of low level of standard requirements, and a poor description of customer's requirements. This problem was considered in current book.

Weiss Technik manufactures battery testing equipment and buttery testing systems that simulate environmental influences under accelerated conditions in corresponding with current standards. The company's test chambers simulate the following conditions:

- *Temperature*: The temperature shock test chamber assesses the effects of extremely rapid temperature changes ranging from −80 °C to +220 °C.
- *Humidity*: The temperature-humidity test chamber can reach 98% humidity (RH) at 95 °C.
- *Tensile strength*: Test equipment can ensure the quality of materials and integrity of design and construction under realistic conditions.
- *Dust*: Equipment injects compressed air through special nozzles and add dust to create a swirl.
- *Water* splash: To determine if housings are protected from water entering.
- *Vibration*: Equipment measures the stress on parts caused by intense vibrations during use.
- *Altitude*: Test chambers can simulate heights of up to 100,000 feet for use of batteries used in aerospace and aviation.

- *Vacuum*: Thermal vacuum chambers accurately reproduce outer space conditions for pressure, cold, solar radiation, UV, and cosmic rays.
- *Stress*: Highly accelerated life testing (HALT) and highly accelerated stress screening (HASS) tests exist in these chambers.

As we can see from the above, the simulation capabilities of these firms are advanced beyond many other companies still meeting current testing standards. Unfortunately, these standards do not offer the possibility for reliability or durability testing, and therefore cannot generate the accurate initial information needed for successful efficiency prediction. Fortunately, the firms reviewed here are exceeding current standards and are providing much of the information needed for high-quality testing.

As was mentioned in this book, using HALT and HASS do not help accurately simulate real-world conditions.

Weiss Technik product includes additional chambers for battery testing: separate dust chambers, corrosion chambers, and spray/splash chambers.

This company also provides testing service: the A2LA Accredited Test Laboratory provides environmental simulation testing utilizing the test technology to meet testing needs from product qualification testing, overflow testing, and/or third-party product validation. Capabilities include temperature, humidity, and/or vibration, thermal shock, burn-in, radiator testing, altitude, vibration, HALT/HASS, shock, salt spray, cyclic corrosion test, and drop testing. Serving provides from two locations in Cincinnati, OH, and Sterling Heights, MI.

The GSPEL, located at the US Army Tank Automotive Research, Development and Engineering Center (TARDEC) in Warren, Michigan, provides the ability to develop, test, and troubleshoot vehicle systems and components under a variety of conditions, leading to more efficient and mobile ground vehicles.

The centerpiece of the GSPEL is the Power and Energy Vehicle Environmental Lab (PEVEL), which provides specialized test chambers. In 2007 the Army contracted Jacobs Engineering (Jacobs) to perform the design/build of the PEVEL facility. This environmental laboratory provides full mission profile testing under various environmental conditions, including temperatures ranging from −60 °F to 160 °F, relative humidity up to 95%, and wind speeds up to 60 miles per hour [45, 46].

In addition to humidity simulation, the test chamber is able to simulate, besides of humidity simulation, various solar loads, with the ability to provide simulation of the sun for the prescribed solar intensity setting of the chamber.

The PEVEL can also simulate road loads for wheeled and tracked vehicles. This can be applied for any situation from a single-drive wheel up to ten-drive wheels. Each wheel load and speed can be controlled independently, enabling full road load simulation.

RENK Systems Corporation [47] has built numerous test stands for heavy-duty and off-road vehicles: trucks, buses, forklifts, other material-handling equipment, and agricultural equipment, including tractors and harvesters. This company has designed two-axle dynamometers for combined testing in climate test chambers.

ESPEC design and manufactures equipment for combined along testing [48] along with mounted equipment usually electronics. While the company inaccurately advertises this as "reliability testing," their work is interesting as it demonstrates a step toward accelerated reliability testing. This company designed and produced equipment for combined multi-axial vibration, temperature, and humidity simulation and testing.

MIRA's [49] combined environment facilities enable tests combining cyclic temperature, humidity, solar loading, and shock/vibration to be run according to a customer's exact requirements. By combining these conditions, real-world environments and damage mechanisms can be reproduced in the laboratory.

A number of facilities are fitted with state-of-the-art fire suppression systems and flexible interfaces designed to provide safe shutdown of equipment based on configurable trigger conditions. These systems facilitate the safe testing of equipment requiring immediate power down, for example, lithium-ion batteries and their management systems.

All of these facilities have access ports to allow power feeds including electrical, hydraulic, and pneumatic connections.

Their available combined environmental testing capabilities include:

- Climatic capability: max temperature ranges, -70 °C to $+180$ °C, and max chamber size, 3 m × 3 m × 4 m
- Humidity range 10–95% RH; solar loading programmable up to 1200 W/m^2
- Vibration capabilities: five EM shaker systems and two multi-axis EH systems; maximum thrust 62kN; frequency ranges: 1 Hz–3 kHz
- Vibration types: sine (including resonance search, track, and dwell); random; shock; sine-on-random, random on random, and gunfire (mixed mode testing); time history replication (road load data, etc.)

Environmental Tectonics Corporation (ETC) [51] Testing and Simulation Systems (TSS) group has been designing and manufacturing environmental simulation systems for the Automotive and HVAC Industries since 1969. ETC's experience in the design, manufacture, installation, and maintenance of environmental simulation systems include expertise in the development of the following:

- Aeromedical (equipment and custom training curriculums for the aeromedical community)
- Automotive testing (drive in test chambers, conditioned air supplier systems, and test rooms)
- Military aviation (ground-based training simulators for a wide variety of tactical training uses)
- Environmental chambers

4.7.4   Combined Testing for Vehicle Components

Link Engineering Company [52] serves the transportation industry in segments such as passenger car, truck, trailer, bus, motorcycle, aircraft, railway, and off-road vehicles. It particularly focuses on providing test solutions for components behind the engine, including transmissions, brakes, clutches, wet and dry friction materials, wheels, tires, hubs, springs, steering systems, axles, and related subsystems. Link also provides a range of electric motor test systems including alternators, starter motors, and steering system testers.

Wheel and Hub Test Systems

Wheel and bearing test stands are available in several system configurations and are capable of testing performance and characteristics through various load inputs. Additionally, auxiliary systems are available for creating different extreme conditions of temperature, humidity, mud, and salt slurry test parameters.

Biaxial Wheel Test System: Designed for advanced design and development testing, full verification, product verification, and gristmill testing of automotive and light truck wheels, and their camber angle control is designed to simulate actual road loads. Their capabilities include:

Radial Fatigue Machine: Tests the performance and characteristics of automotive wheels through a tire interface and road wheel system

Wheel Impact Test Stand: Provides a means of simulating aggressive curb impacts on wheel and hub assemblies

Bearing Spalling Test System: Provides both radial and axial loaded testing of hub bearings

Transmission and Driveline Test Systems

As a multifaceted engineering developer and comprehensive manufacturer, Link provides fully integrated test systems for the evaluation of transmissions and driveline assemblies and components. Their equipment works with transmission and driveline systems including oil and friction, reaction plates, gear interface, belt and chain, housing, torque converter, piston, splined shaft, output shafts, axles, differentials, wheel ends, mounting hardware, full and partial systems, and assemblies.

Examples of the transmission and driveline test systems that Link Engineering develops and manufactures include:

- High-speed automatic transmission test system
- SAE No. 2 Wet Friction Test Stand
- Manual Clutch Durability Test System
- Transmission Torque Cycling Durability System
- T0-4 Test System
- Four Square Test System

Cincinnati Sub-Zero's AV/CV-Series vibration chambers offer rapid temperature change rates with combined temperature, humidity, and/or vibration [53]. These chambers may be used with existing vibration shaker tables and can also be utilized as separate temperature/humidity cycling chamber providing a greater return on their investment.

Their drive-in chambers simulate vibration and a variety of climate conditions including ultraviolet simulation and infrared lighting.

Their CV-Series temperature/humidity and vibration chambers are designed to integrate with vibration systems only in the vertical mode of operation. Chambers may also be used as temperature cycling chambers without vibration utilizing a solid floor plug.

Cincinnati Sub-Zero has also developed a drive-in test chamber for testing whole vehicles. The BMW Climatic Test Complex [54] consists of three large climatic wind tunnels, two smaller test chambers, nine soak rooms, and the support infrastructure.

The capabilities of the wind tunnels and chambers are varied, and on the whole give BMW the ability to test at practically all conditions experienced by their vehicles, worldwide. The wind tunnel test section was designed to meet demanding aerodynamic specifications, including a limit on the axial static pressure gradient and low-frequency static pressure fluctuations.

As mentioned earlier, many combined testing technologies do not offer the possibility for successful product efficiency prediction. The basic cause of this shortfall is an erroneous strategy of simulating fewer than three elements from the many influences real-world conditions in one test chamber or proving grounds. This contradicts real-life situation, where many components of this situation act simultaneously and in combination (see Refs. [1–3]) and others.

This wrong strategy continues currently and has trends in the near future.

4.7.5 Equipment for Accelerated Reliability and Durability Testing

Figure 4.20 illustrates a test chamber with computer-controlled universal equipment for different types of engine reliability/durability testing that includes [1, 2]:

- Temperature
- Humidity
- Vibration
- Dynamometer
- Pollution chemical
- Pollution mechanical
- Input voltage

One specific difference of this testing equipment, as compared with the equipment cited below, is that it is used for engine testing by companies that design and manufacture the engines. It is when this equipment uses vibration testing in

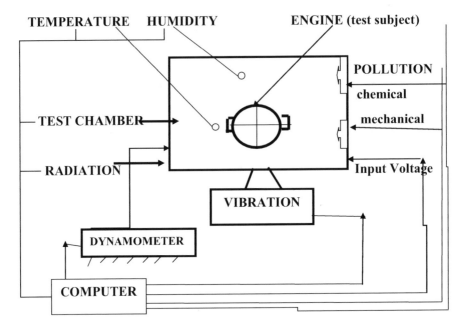

Fig. 4.20 The scheme of equipment *in test chamber* for engine reliability/durability testing

conjunction with pollution (chemical and mechanical), full spectra of radiation, temperature, humidity, and input voltage that a more accurate real-world simulation for reliability and durability accelerated testing.

A similar test chamber to that shown in the above scheme was developed by the State Enterprise TESTMASH for transmission's reliability/durability testing. This author participated in the development of a group of test chambers for reliability/durability test equipment that combined specific vibration equipment (in three and six axes), equipment for technological process simulation, drive simulation, and equipment for mechanical and corrosion process simulation. All of the loadings applied to the test subject were random processes so as to be close to real-world conditions.

TESTMASH developed also a drive-in test chamber with universal equipment for accelerated reliability and durability testing [55].

As depicted in Fig. 4.21, this chamber simulates the following fully integrated input influences with modern system of control:

- Vibration testing equipment with random process simulation
- Dynamometer testing equipment
- Equipment for simulating chemical air pollution
- Equipment for simulating mechanical (dust) air pollution
- Equipment for simulating full range of solar radiation
- Equipment for simulating input voltage
- Equipment for simulating various temperatures
- Equipment for humidity simulation

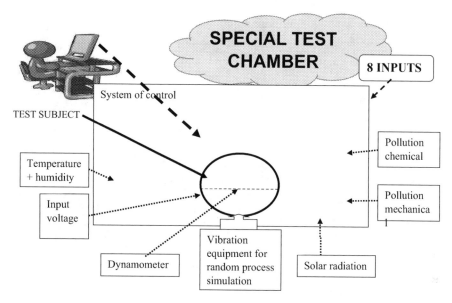

Fig. 4.21 Scheme of the test chamber for accelerated reliability/durability testing (ART/ADT) (TESTMASH) (8 inputs)

Figures 4.22 and 4.23 demonstrated the components of the test chamber (Fig. 4.21) for random vibration testing of wheeled product in three degrees of freedom. One can see in front cover of the book [3A] the TESTMASH's design of random vibration equipment in six degrees of freedom.

The Automotive Centre of Excellence (ACE) Institute of Technology at the University of Ontario at the beginning of the twenty-first century created, acquired, and operated a climatic wind tunnel. This tunnel was designed to provide automotive manufacturers, tier, suppliers, and other industries of all sizes with testing capabilities to validate prototype products under a range of simulated field conditions and that could be used in the future, after necessary development, for accelerated reliability/durability testing [56].

The project in which ACE developed a testing facility in partnership with the government and a university provided solutions to many interrelated problems such as cost, utilization, and availability of testing facilities for accelerated reliability and durability testing. However, such a project requires strategic long-range thinking, cooperation of varied partners, and the commitment of resources by all concerned parties.

The solution developed by ACE followed the ideas, presented in this book, and can be useful for future implementations of these ideas by others.

In continuation of the above system, multiple chambers in one system (Fig. 4.24), which can significantly reduce the cost of accelerated reliability/durability testing, were developed. The advantages of such system, in comparison with using the abovementioned chambers for testing one test subject (vehicle or its component), include:

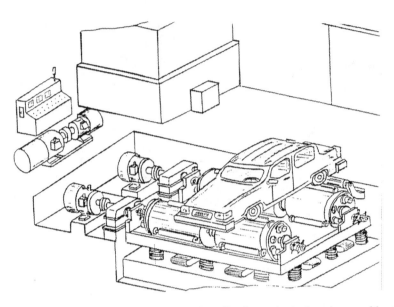

Fig. 4.22 Scheme of test equipment for car's random vibration testing in three degrees of freedom

Fig. 4.23 Test equipment—component of test chamber (Fig. 4.21) (TESTMASH) for physical simulation random process of vibration in three degrees of freedom

- Ability to test simultaneously multiple test subjects in one system.
- A single power system for N test chambers.
- A single computerized data acquisition and control system.
- A single exhaust system for testing multiple test subjects.
- A single system of input influence simulations for N test chambers.
- Compared with a traditional block-built test center building, modular test facility solutions save a lot of time and trouble.

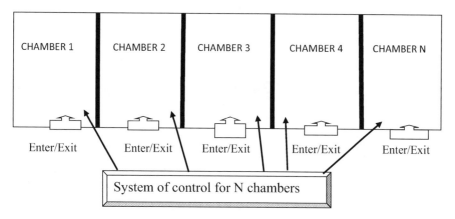

Fig. 4.24 A common scheme of modular test center in one system (with one combined system of control)

A brief history of Automotive Centre of Excellence (ACE) will be provided. The ACE was developed in partnership with the University of Ontario. This innovative government-industry-university partnership is extraordinary in its scope and vision and remarkable for its inclusiveness. Its founding partners have created a welcoming environment where the dynamic pursuit of automotive and manufacturing innovation can involve everyone with a great idea, a desire to learn, and a commitment to sustainable progress. As the ACE project evolved, the university partners committed over $100 million including contributions from the Government of Ontario, Government of Canada, UOIT, and General Motors of Canada through its PACE partnership.

The design and engineering work started in 2007, with construction beginning on 2008. Five different countries were involved in the construction of the ACE facility.

The CRE offers full-size chambers that allow for full climatic and life cycle, including one of the largest and most sophisticated climatic wind tunnels. In this test chamber, wind speeds can exceed 240 kilometer per hour, temperatures range from −40 to +60 °C, and relative humidity can range from 5% to 95%. The climatic wind tunnel has a unique variable nozzle that can optimize the airflow from 7 to 13 square meters (and larger) allowing for an unprecedented range of vehicle and test property sizes (see component of this tunnel in Fig. 4.25). Coupled with this feature is a large flexible chassis dynamometer that is integrated into an 11.5-m turntable.

Now, for the first time anywhere, vehicles and test properties can be turned into the airstream under full operating conditions to facilitate analysis in crosswind conditions. The large open chamber has a readily reconfigurable solar array that will replicate the effects of the sun. Also it is hydrogen-capable, allowing for alternative fuels and fuel cell development. Features of the climatic wind tunnel will include a 7–14.5 m^2 variable nozzle concept to enable a wide range of vehicle sizes and wind speed combinations. A turntable with a chassis dynamometer and vibration

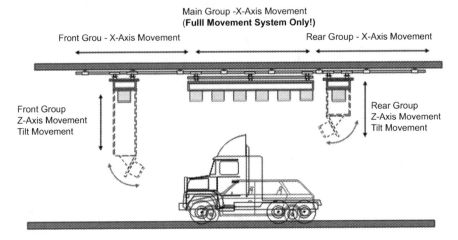

Fig. 4.25 Test section with solar simulation system (A & B) [56]

equipment enables vehicle yaw tests. The dynamometer assembly has been designed for removal from the test chamber and to include an elevator assembly and air bearing transport system.

Here are a few of the ACE Climatic wind tunnel highlights:

- Climatic control
- Dimensions: L20.1 m × W13.5 m × H7.5 m
- Diurnal solar array
- Fuel dispensing (hydrogen-capable)
- Snow, rain, and arid desert
- Temperature and humidity (60 °C to −40 °C, 5–95% humidity)
- Turntable with a chassis dynamometer to enable vehicle tests at yaw
- Variable nozzle (7–14.5 m²) to enable a wide range of vehicle sizes and wind speed combinations
- Wind speeds exceeding 240 kilometers per hour (km/h).

Fan and motor system:

- Airflow—aero 888 m³/s
- Airflow—large/small nozzle 293/470 m³/s
- Expansion/vibration isolation
- Motor speed 610 r/min
- Nominal power 2500 kilowatts (kW)
- Rotor/blade mass 6000 kilograms (kg)
- 0–40% in 6 s, 40–100% in 15 s, 100–0% 15 s

The air circuit viewing platform (ground floor) is illustrated here.

For accelerated testing, especially in space problems, the problem of sensors remains very important. One example presented in [17] shows how the operating characteristics of interest are included.

References

1. Lev M. Klyatis, Eugene L. Klyatis (2006) Accelerated Quality and Reliability Solutions. Elsevier.
2. Lev M. Klyatis (2012) Accelerated Reliability and Durability Testing Technology. Wiley.
3. Lev Klyatis (2016) Successful Prediction of Product Performance: quality, reliability, durability, safety, maintainability, life-cycle cost, profit, and other components. SAE International.
3A. Lev Klyatis (2020) Trends in Development of Accelerated Testing for Automotive and Aerospace Engineering. Academic Press (ELSEVIER).
4. Honda civic fleet and accelerated reliability testing (2005).INL/EXT 06-01262. Energy efficiency and renewable energy. U.S. Department of Energy.
5. Birch S. (2007) 24 million km of testing for Mercedes-C Class. Automotive Engineering.
6. LMS Supports Ford Otosan in Developing Accelerated Durability Testing Cycles," http://www.lmsintl.com/LMSFord-Otosan-developing accelerated durability testing cycles.
7. Dynamic Research, Inc. (DRI).
8. The Promise of Virtual Testing. MTS Systems Inc.
9. Reilly, T. (2018) "Satellite and Spacecraft Vibration Testing Control," Aerospace Testing International, SnowCase,
10. Fatemi, S.Z., Guerin, F., and Saintis, L. (2012) "Development of Optimal Accelerated Test Plan," in *RAMS Proceedings.*
11. Horiba, MIRA, Vehicle Durability Testing.
12. Schenkeiberg F. (2012) Determine and design the best ALT. RAMS Proceedings.
13. Crashworthiness. Wikipedia.
14. Insurance Institute for Highway Safety. Highway Loss Data Institute. Overview.
15. Simon Parkin (2016) Crash test geniuses. Crash Test Technology International. September
16. https://media.daimler.com/marsMediaSite/.../Crash-testing-for-safety-research.xhtml.
17. Frank Murrau, S. Hestmat, H. Fusaro (1995) Accelerated Testing of Space Mechanisms. April 1995. MTI Report 95 TR 29.
18. Simon Edmonts (2016) Quarcycle testing. Crash Test Technology International. September
19. Bruce Coons (2018) A Secret Weapon for Roof-Crush Testing. SAE Automotive Engineering. May
20. Lev M. Klyatis, Edward L. Anderson (2018) Reliability Prediction and Testing Textbook. Wiley.
21. Antony James (2013) Hit The Dust. Aerospace Testing International. September
22. Jessy Cavazos, Industry Director. Frost & Sullivan. www.frost.com
23. Hobbs, G.K. (2000) Accelerated Reliability Engineering: HALT and HASS. Wiley,.
24. Paul Seredynski (2021) New PDE test chamber. Automotive Engineering. SAE. May
25. Lev Klyatis. Analysis of Current Trends in Development Accelerated Testing. SAE 2022 World Congress. Paper # 2022-01-0212
26. P. T. Parker, H. A. Chan (1999) Product Reliability Through Stress Testing. Annual Reliability and Maintainability Symposium Tutorial Notes.
27. Lev Klyatis, Lesley Walls (2004) A Methodology for Selecting Representative Input Regions for Accelerated Testing. Quality Engineering, Volume 16, Number 3, 2004.
28. Lourie A.B. (1970) Statistical Dynamics of Farm Machinery. Leningrad. Kolos.
29. J. A. Corn (1996) Random Processes Simulation and Measurement. McGraw-Hill. New York.
30. F. Nome, G. Hariman, L. Sheftlevich (2007) The Challenge of Pre-Biased Loads and the Definition of a New Operating Mode for DC-DC Converters. Power Electronics Specialists Conference. 2007. IEEE.
31. Robert Harwood (2012) The Role of Engineering Simulation in the Evolution of Unmanned Aircraft Systems. Engineering Solutions for Military and Aerospace. DEFENSE Tech Briefs. SAE International. Supplinaire to NASA Tech Briefs. Volume 8. December 2012.
32. B. I. Brecher (2000) Accelerated Testing Experience with Avionics. The 54th Annual Quality Congress Proceedings. May 8–10, 2000. Indianapolis.

33. Business Wire (2017) A Berkshire Hathaway Company. https://www.businesswire.com/... Top-3-Emerging-Trends-Impacting-Global-General. Graphic: Business Wire, May 17, 2017.
34. Electronic Test and Measurement Market (2018) Global Trends, Market Share, Industry Size, Growth, Opportunities and Forecast to 2023. "WiseGuyReports.com adds "Electronic Test and Measurement Market 2018 Global Analysis, Growth, Trends and Opportunities Research Report Forecasting to 2023".Tuesday, October 16th 2018.
35. Automotive Test Equipment Market (2018) Global Emerging Technologies, Top Key Leaders, Recent Trends, Industry Growth, Size and Segments by Forecast to 2022. Posted: Sep 12, 2018
36. Automotive Test Equipment Market Overview (2018) New York, United States – October 18, 2018 /MarketersMedia
37. Charles Sweetser, OMICRON, www.omicron.at
38. Mark Schrepferman, Peregrine Semiconductor Corporation, www.psemi.com
39. Michael Fox. Conception and build-out of the FLIP Test and Measurement Product Line. RPX Technologies, Inc. May 2016
40. Burke Porter Machinery Company. www.bepco.com
41. Test Centre – European Test Services (ETS). www.european-services.net
42. European Space Agency. PepiColombo. Mercury Composite Spacecraft. Sci.esa. int/.../50547 – sunshield-being-installed-on
43. ETS LINGREN (2014) Test and Measurement Chambers & Equipment. EE. Evaluation Engineering October 2014.
44. Automotive Chambers. ETS Lindgren. www.ets.-lindgren.com
45. Ground Systems Power and Energy Laboratory – U.S. Army. www.army.mil/standto/archive/issue.php.issue...04
46. Jacobs (2012) Environmental testing for military vehicles. Automotive Testing Technology International. November 2012, pp. 78–79.
47. RENK Systems Corporation. www.renksystems.com/
48. ESPEC Technology Report (1997) Special Issue: Evaluating Reliability.No. 3. 1997.
49. Combined Environment Testing | MIRA. www.mira.co.uk › *Defense Vehicle Engineer.*
50. Weiss Technik. Safe Automotive Battery Testing with Environmental Test Chambers. Environmental Test Chambers Solutions EV/Battery. Weiss <partner_maiullings@ukip-artner.com.
51. Environmental Tectonics Corporation (ETC). https://www.etcusa.com/
52. Link Engineering Co. www.linkeng.com
53. Cincinnati Sub-Zero Chambers. All types and sizes made for you. www.cszindustrial.com/Environmental
54. Bender T., Hoff P., Kleemann R. (2011) The New BMW Climatic Testing Complex – The Energy and Environment Test Centre. Paper 2011-01-0167. 2011. Detroit.
55. High Quality Equipment for Test Service of Farm Machinery (1990) Interview with Chairman of Engineering Center TESTMASH Dr. Lev Klyatis. Journal Tractors and Agricultural Machines. No. 11, 1990. Moscow, Soviet Union.
56. UOIT – ACE – Automotive Centre of Excellence. ace.uoit.ca/

Chapter 5
Negative Trends in the Development of Simulation, Testing, and Prediction in Engineering

5.1 Introduction

As one can see from this book, the simulations connected closely with testing, as testing is connected closely with prediction. Therefore, the level of simulation being done has an influence on the level of testing, and the level of testing influences the accuracy of predictions. This especially relates to the simulation of real-world conditions. Commonly, the accuracy of simulating of real-world conditions for research and testing is the key factor for obtaining successful results from research, testing, and prediction. Finally, the level of prediction technology directly influences the improvement of engineering product efficiency.

Basic positive and negative trends in the development simulation, prediction, and accelerated testing are described in detail in [1–3], as well as other author's books. Here one can see basic negative trends in the development of the above factors.

Usually, there are both positive and negative aspects of each innovative technology or product development. If the positive aspects (quantitatively and qualitatively) of a modern technology or product are superior to those of competitors, this innovative technology or product is adopted. If not and it is a net negative, its success is doubtful as well as its continued usage.

There are many publications about positive trends of prediction, simulation, and testing with far fewer publications about negative trends. Therefore, the author will concentrate his attention below on negative trends in simulation development, accelerated testing and prediction in engineering. This focus is very important, because negative trends in this area are leading to decreased quality, reliability, durability, maintainability, supportability, safety, and economic aspects of developed products.

Supplementary Information: The online version contains supplementary material available at https://doi.org/10.1007/978-3-031-16655-6_5.

As was discussed earlier, product recalls have been increasing over the years. This phenomenon reflects concurrent trends in decreased reliability, maintainability, safety, profit, increased fatalities and injuries, and higher life cycle costs (see Introduction). The data being published about recalls has been proven to be very accurate. Therefore, it can be used for an objective analysis about disturbing trends for improving engineering product efficiency.

The number of deaths from incidents is another very important statistic, because of its tragic circumstances. Whatever reduces these numbers is obviously an important engineering concern. Most fatal accidents are generated by human error such as speeding, reckless driving, or drunken accident. While human error cannot be mitigated, protecting the occupants in cars during impact does become an engineering concern. Any failure of protection indicates an inaccurate prediction of the car's capability to withstand impact damages. Of course, improved design of vehicles, especially system of control, can improve the possibility for the accidents, which leads to another inherent importance of successful prediction.

Poor efficiency prediction is usually the failure to adopt effective prediction methodology and accelerated testing development methodologies or at best a very slow adoption of them. These sad trends can be easily seen with a cursory review of most published books, journal and magazines articles, and conferences proceedings around the world. Most of them discuss product engineering designs with little commentary given to the efficiency and quality of these products. This paucity reflects the misguided attention of researchers, designers, and producers alike. As a result, the predicted quality and efficiency of products do not correspond to what the user purchases.

As was demonstrated in [23], engineers often use poor testing and prediction methodologies reflected in the literature and do not understand the consequences of these directions. Below are several examples of poor guidelines being published:

- DoD 3235, 1 H. *Test and Evaluation of Systems RAM* [33] has the following negative aspects:

  - Advocates poor testing technique
  - Advocates poor testing strategies that do not includes the real-world simulation
  - Does not consider the real reasons of degradation and failures
  - Does not include how one can design accelerated testing and quickly evaluate the product reliability and quality to eliminate poor prediction techniques
  - How one can determine a product's efficiency in the laboratory that reflects its field efficiency

- IEEE 1413-2002: *A Standard Methodology for Reliability Predictions: Guide for Selecting and Using Reliability Predictions based on IEEE 1413. 2019* [34]

This document is intended to provide a framework for reliability prediction procedures for electronic equipment at all levels. The updated IEEE Standard 1413 Methodology for Reliability Predictions and Assessment for Electronic Systems and Equipment in reality as follows:

- Consists of mostly reliability prediction theory and is only mentioned in passing the testing processes for obtaining initial information for accurate prediction. Therefore, it cannot be useful for practical reliability prediction.
- Based on mostly theoretical statistical approaches to prediction and testing.
- Bases its test data predictions on generic testing methods without considering what sort of specific testing is needed for problems at hand. Therefore, most of them are not appropriate for successful reliability prediction.
- Mean generic methodologies are useful only for software prediction during any predesign phase, but not for prediction of physical product reliability during design and manufacturing stages.
- Is based mostly on the literature published in the twentieth century that stagnates the development of better prediction testing processes.
- Leads to inaccurate practical prediction, because it is not based on accurate simulation of real-world conditions.

When considered predictions are based on field data, the actual efficiency of an item reflects accurately its actual operational environment. Thus, its components' efficiency predictions based on field data are appropriate for an item already in service (e.g., for logistics planning, warranty reserve, repair department sizing, or future corrective action). Field data are also used when comparing reliability predictions based on the actual test data or analysis and the actual efficiency of the equipment for a brief time. The type and quality of field data can range from simple factory shipping and return data to sophisticated tracking of installation times, operating hours, and failure times for every unit in service. When the field reliability data is used for an item's reliability prediction, testers are not taking into account how this needs many years of recording a product's service life.

Accelerated stress testing in the field does not take into account some basic degradation processes like corrosion or changing usage technologies.

- MIL-HDBK-217F, *Military Handbook: Reliability Prediction of Electronic Equipment* [35]

  - The impact of proposed design changes on reliability can be determined *only* by comparing the reliability predictions of the existing and proposed designs.
  - There are several important limitations of this document's technologies:

    The failure rate model uses point estimates that are based on available data. Hence, they are valid for the conditions under which the data were obtained and for the devices covered. Some extrapolation during model development is possible, but their applications to the inherently empirical nature of the models can be severely restrictive. For example, none of the models in this handbook predict nuclear survivability or the effects of ionizing radiation.

    Another limitation of its considered prediction methods are the mechanics of the testing processes. Here is a typical problem. The part stress analysis method requires a significant amount of design detail. This naturally imposes a time and cost penalty. More significantly, many of the details are not available in the early design stages. For this reason, the handbook contains both the part stress analysis method and a simpler parts count method that lead to inaccurate testing results.

A basic limitation in its approach to reliability prediction is its dependence on correct applications by the user. Those who view this prediction only as a number that must exceed a specified value can usually find a way to achieve their goal without any impact on the system. Whether this concentration on not impacting a system result in accurate test data is consequently problematic.

- Military Standard MIL-STD-781D *Reliability Testing for Engineering Development, Qualification, and Production* [36]

  - Consists mostly of definitions, general requirements, brief detail requirements (one page), notes, and tasks (test planning and control, integrated reliability test plan document, reliability test procedure, reliability growth planning, test program reviews, joint test group, reliability test reports, development testing, reliability accounting tests, and environmental test screening).
  - Doesn't include accurate simulation of field conditions as a basis for efficiency prediction. The above sources and other literature, for example [11, 12], demonstrate only that there is a problem with the implementation of successful efficiency or reliability prediction solutions.

As was written in [12]:

Inaccurate reliability predictions could lead to disasters such as in the case of the US Space Shuttle failure. The question is: what is wrong with the existing reliability prediction methods? This paper examines the methods for predicting reliability of electronics. Based on information in the literature, the measured vs predicted reliability could be as far apart as five to twenty times. Reliability calculated using the five most commonly used handbooks showed that there could be a 100 times variation. The root cause for the prediction inaccuracy is that many of the first-order effect factors are not explicitly included in the prediction methods. These factors include thermal cycling, temperature change rate, mechanical shock, vibration, power on/off, supplier quality difference, reliability improvement with respect to calendar years and aging. As indicated in the data provided in this paper any one of these factors neglected could cause a variation in the predicted reliability by several times. The reliability vs ageing-hour curve showed that there was a 10 time change in reliability from 1000 ageing-hours to 10,000 aging-hours. Therefore, in order to increase the accuracy of reliability prediction the factors must be incorporated into the prediction methods.

Standards in FMEA consider only individual components of product and relates not only to efficiency but also their components as reliability, durability, and safety and therefore not consider in the totality of its real-life environment, where all efficiency components interact.

The author's experience in standardization, including preparation of ISO and IEC international standards, as well as SAE (Society of Automotive Engineers), ASAE (American Society Agricultural Engineers), and ASQ (American Society for Quality), opens the possibility that during the formulation of these standards preparation, people often demonstrated narrow thinking and did not take into account the real-world situation. For example, when the author, during his work with the IEC/ISO Joint Study Group in Safety Aspects of Risk Assessment, proposed the inclusion of biological factors as components of risk and safety engineering, it was not accepted. The basic reason given was that biological factors were not seen as being important factors for engineering product safety and risk. One can see the author's analysis about this in Chap. 6 (Fig. 6.17) of this book. But consideration of these factors is very important. Another example was with SAE G-11 Committee's not approving the group of draft standards in reliability testing. And during 20 last years, the standard reliability testing was in SAE G-11 Fact Sheet without solution.

This situation continues until now. Specifically, current international symposiums and conferences on testing consider mostly ground, flight, or laboratory testing of vehicles without offering any information on successful efficiency prediction, but rather for reliability and safety prediction of product also.

For example, the European Test and Telemetry Conference (ETTC'21), June 15–16, 2021, considered only the technological aspects of flight and ground testing for improving performance by pilots and drivers. As a result, standardized tests do not help enough to improve a product reliability, quality, safety, durability, life cycle cost, and other components of product efficiency improvement during service life.

If we review the proceedings of other symposiums and conference programs, we will see how laboratory tests being conducted are based on inaccurate simulation of real-world conditions that do not account for the interconnection between different input influences, influences with human factor and safety. Instead, they concentrate on the testing of separate product components (details and units).

Therefore, their predictions of efficiency or any of their components are unsuccessful.

There are several actions that can be employed to eliminate many of the reliability, durability, and safety problems found in engineering, including several types of testing whose objectives should be as follows:

- Finding the real reason, not **only** the result of the problem
- Studying the real correct reason for the problem
- Deciding on how one can eliminate or, as a minimum, mitigate to an acceptable level the reasons that are creating the problem
- Developing and implementing the recommendations for new procedures and using the more effective knowledge about elimination of the above reasons through advanced methods of testing level needed for assuring product efficiency
- More use of accelerated reliability and durability testing, especially for new development processes such as autonomous vehicles (AV) and processes digitalization

Unfortunately, the testing methodologies and equipment, as well as prediction methodologies, are often directed at making them cheaper and easier to conduct. Consequently, their negative results are not found during testing but later with often tragic results. Examples of these actions and results using can be found in [2].

Recently, there have been some worldwide publications, including the area of autonomous vehicle testing that have provided relevant information that have led to positive results. For example, the referenced publication [2] provided different ideas for improving the current status of accelerated testing. Elsewhere, for example, in [4], the authors briefly reviewed their concepts for accelerated testing. They also gave an overview of some of their current and planned research to improve accelerated testing planning and methods. Otherwise, sadly, the agendas of recent North America and Europe symposiums and conferences on autonomous vehicle test and development that were nominally devoted to testing did not list a single title addressing new trends in testing developments. This situation reflects a common lack of appreciation about the importance of testing.

5.2   Some of the Basic Negative Trends in the Technology Development of Simulation and Accelerated Testing

From author's previous publications [1, 2, 4], it can see how accelerated testing effectiveness and successfulness prediction of product's efficiency consists of several basic steps (Fig. 5.1):

But these steps are not considered in most of used approaches. For analysis of the negative trends in technology development of real-world simulation and accelerated testing, let us consider the latest literature that is specialized in automotive and aerospace technologies.

As was written in Chap. 4 of this book, "... combined testing do not offer the possibility for successful product efficiency prediction. The basic cause is wrong strategy of simulation separately not more than two-three from many components of real-world conditions in one test chamber or proving grounds. This contradicts real-life situation, where many components of this situation act simultaneously and in combination (see [4.1], [4.2], [4.3]) and others. This wrong strategy continues currently and has trends in near future."

There are two basic groups of causes responsible for the negative aspects occurring in simulation, accelerated testing, and efficiency prediction development—mentality (psychological) and engineering. Their brief description can be seen in Fig. 5.2. These negative aspects lead to the two basic causes of recalls, which are

Fig. 5.1 The basic steps of accelerated testing effectiveness and successful efficiency prediction of the product

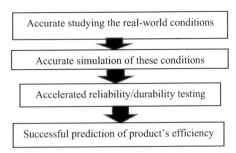

MENTALITY (PSYCHOLOGICAL)	ENGINEERING
• Misguided management focused on testing costs rather than benefits. • Professionals believing that real-world situation can be accurately simulated in the complex of laboratory work and periodical field testing. • Therefore, they use simpler and cheaper simulations that are inaccurate. • Misunderstanding the role of interconnectivity between input influences on the product's reliability, durability, and other efficiency components, and therefore simulate only separate input influences or their several components. • Confusing about two important concepts – cause and result. • Inaccurate use of terms and definitions. For example, they provide vibration testing, but call it durability testing. • Not clear understanding that current level of virtual testing is not based on accurate simulation the entire complex of real-world conditions.	. • Undeveloped methodology for accurate simulation the field conditions. •The lack of accelerated stress testing theory. • Not seeing the important role of accurate field simulation in development testing processes and product efficiency. • Using inaccurate strategies for simulation of real-world conditions. • Using inaccurate standards found in literature. • Using only AV technological (driving) testing instead of testing them in combination with reliability/durability testing. • Undeveloped equipment for accurate simulation of real-world conditions. • Undeveloped digitalization technology for accurate simulation of real-world conditions. • Unsuccessful testing of artificial intelligence concepts from using the above negative aspects. • Using simulation separate or partly input influences instead of real-world full range simulation

Fig. 5.2 Brief depiction of two basic groups that cause negative aspects on accelerated testing and successful prediction development

using the negative aspects of simulating real-world conditions, that lead to inaccurate prediction. This can be seen in Fig. 5.2.

The result of these causes are inaccurate simulation of real-world conditions leading to poor methodologies and equipment for accelerated testing and then to unsuccessful prediction of product efficiency.

The next result is lower than predicted safety, reliability, durability, and higher life cycle cost.

The final results are increased death, recalls, and economic losses (Fig. 5.3).

Mentality Causes

In more details:

- Most professionals, especially senior managers who are concerned about short-term investment returns, erroneously think that using simpler less expensive accelerated testing that has been called accelerated life testing (ALT) is adequate. Hence, they are reluctant to invest in more complex testing even though their produces are far more complex and need testing resources to much. They think this because they believe testing to be a separate and not an expensive process. They do not take into account the subsequent incurred costs and procedures during design, manufacturing, and usage that are direct result of failed predictions.

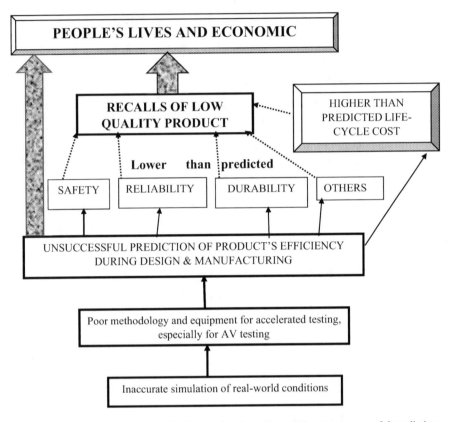

Fig. 5.3 The path from inaccurate simulation of real-world conditions to unsuccessful prediction, recalls, people's life, and economic losses through lower than predicted safety, reliability, durability, maintainability, and other efficiency components

For example, simple vibration testing is far less expensive than accelerated reliability and durability testing (ART/ADT) with an accompanying full simulation of the real-world conditions. But vibration testing alone is based on simulations of one or two elements and ignores all other interactive elements of the product's field conditions. Consequently, such testing the accurate results is needed for studying the product's full real-world performance. The net result is insufficient performance and disappointed customers.

- Often managers cannot believe that actual field situations can be accurately simulated in the laboratory coupled with periodic field testing. But as it turns out, simulations do portrayal of reality when properly performed. Through accurate multi-influence simulations, one can study most of the components of nature acting on a product with faster results than conducting field/flight testing. Accurate simulation is achieved by transferring the flight/field influences to the laboratory with a high degree of accuracy. (Of course, even with this some field-/flight-specific influences may need to be investigated as an additional testing component). Thus, using ART/ADT through accurate simulations provides a shorter testing timeline and reduced costs for a complete life cycle of design through manufacturing and usage.

- People often confuse causes and results. Frequently, "experts" will state that safety and reliability problems are the causes for recalls. But, in fact, safety and reliability problems are not the causes of the recalls, fatalities, and economic losses. They are the result of failures in the prediction of product's efficiency. Had correct testing procedures been followed, recall rates would plummet.

- The current capabilities of virtual simulation cannot accurately reflect the entire complex of real-world conditions. Therefore, the failures to predict real-world operational failures properly during the design and manufacturing phases are the real causes of the recalls, fatalities, and economic losses (Fig. 5.3).

Shown below are the basic reasons why recalls are increasing from year to year (the number of recalls from 2000 to 2019 have increased more than two times as can be seen in Introduction, Table 1).

- A basic negative influence that slows the implementation of advanced accelerated testing is the mindset by many process managers, engineers, and academia, that minimizing the cost and simplifying testing are an organization's primary goals. Often this results in testing for testing's sake alone, rather than the product's effective development. This approach does not take into account important real-world realities such as the following:

  - More complex products necessitate more complex test methods and test equipment for the accurate simulation of real-world conditions needed for successful accelerated testing.
  - The failure to identify the true degradation mechanism through proper testing influences all subsequent steps of design, manufacturing, and usage.
  - The costs associated with failures resulting from inaccurate prediction are rarely accounted for. The myriad of other design, manufacturing, and usage problems that are directly related to failed testing are rarely considered or quantified. Only direct cost and time are considered. While this viewpoint reduced initial testing costs it ultimately increases costs by generating unscheduled improvements in design and manufacturing processes that remediate customer complaints or product recalls. This nearsighted focus on immediate costs detract from the ability of upper managers to see and appreciate the need for reducing the total life cycle costs that dwarf immediate testing costs.
  - Reducing development times from design to market by relying on inaccurate simulation and testing processes.
  - Using the expression "real world" to describe testing the protocols that in many cases are far from reflecting real-world conditions or operations. One can see the real-world conditions in Figs. 3.2–3.6, Chap. 3.

- Using old approaches such as the Monte Carlo or exponential distribution for approximations, that do not accurately reflect the nonstationary random processes of real life.

The author has demonstrated examples of these in his books [1–3].

As was mentioned, there are strategic and tactical negative aspects of trends in development simulation and accelerated testing.

Strategic negative aspects include a wide range of the following in engineering practice:

- Increasing usage of virtual accelerated testing that relies on the use of computer simulations for physically natural product. As is known, computer simulations are not useful for complete cars, trucks, aircraft, satellites, and research stations being sent to other planets. The basic reasons for this increased usage of virtual (computer) simulations is the mistaken perception of being less expensive. However, as already discussed at length, the quality of simulation for testing is less accurate and more costly over the long term.
- Less attention to development of physical simulation of real-world conditions, because this is complicated and more expensive than virtual simulation.
- Strategic negative trends in the development of simulation and accelerated testing have a global character and are not dependent on the type of testing machines or its units. Strategic negative aspects for accelerated testing include a wide range of the following factors commonly found in poor engineering practices.

  - Excessively rapid implementation of virtual accelerated testing that is leading to using more computer simulation of physical products. As has been shown, computer simulation is not yet ready for real-world accurate simulation, and therefore not useful, for complex products such as complete cars, trucks, aircraft, satellites, and space research devices for interplanetary exploration. While virtual (computer) simulation appears to lead to a savings in time-to-market reduced costs of testing, costs of poor quality (accuracy) results is higher.

- Compared to advanced types of physical simulation and testing, the accuracy of computer simulation for complex products is often less.

  - Real-world conditions are complicated and more expensive to duplicate than virtual simulations. When reviewing these costs, management looks upon testing processes as nonrevenue generating losses without taking into account the financial life cycle failures generated by inaccurate simulation.
  - A wide-range practice of using the wrong definitions associated with simulation and accelerated testing, fatigue testing, accelerated reliability testing, reliability testing, accelerated durability testing, durability testing, proving ground testing, vibration testing, and corrosion testing.
  - The increasing gap between the technical progress in the design and manufacturing of new products and simulation and accelerated testing, because there

is less attention focused on the development of the physical testing requirements.

– Narrow thinking about the need for and the value of accelerated testing as key component of a new design's efficiency. This was described in detail in this author's previous publications [1, 2, 6].

– Status quo thinking in not wanting to initiate the institutional challenges and work involved in developing new physical testing approaches. Why should you be the one to rock the boat? If any previous way worked before, so why change it?

Moreover:

The speed and complexity of technical progress in the design and manufacturing of new products is rapidly increasing. We see this especially in the design and manufacturing development of new product in the electronics, automotive and aerospace areas.

But, at the same time, the speed of technical progress in accelerated testing (testing) technologies has been advancing very slowly as compared to the technical progress in the design and manufacturing. This is easily demonstrated by looking at the developments in both design and testing technologies for the last dozen years (Fig. 5.3).

Further the "modernized" types of testing such as HALT (highly accelerated life testing), HASS (highly accelerated stress screening), AA, and other testing techniques are only a combination of some of the separate influences encountered in real life that involve many influences. Often in using these methods, the loads and the influences used are greater than the maximums of those experienced by actual field loads, and these assumptions change the physics of degradation process. This means that the development of testing needs to keep pace with and reflect the advances in product design and development [1]. Simply stated, more complicated design brings about the need for more careful and complicated testing.

Sadly, in these latter years, the technical progress in testing area is going down (Fig. 5.3). It is connected with increased usage of software simulation and decreased usage of physical simulation. But when reading recent literature, we will often see the terms "revolution" and "revolutionary" being applied to solutions to simulation and testing area. The author is recommending that readers must be careful when seeing these words describing simulation and testing advances.

One example follows in the article "Making the Strategic Move into AV Testing" [5] that lauded "a critical leap into automated vehicle testing." In fact, road testing was cited, which is a traditional technological testing. Nothing was written about reliability, durability, maintainability, or supportability testing. The article further stated: "Despite the increased fidelity of the latest "sim" tools, FEV acknowledges up-front simulation and physical road testing are equally needed at this stage of the automated vehicle era."

And:

"While power train testing and validation are performed within a robust set of established standards, the same is not yet true for AV testing." This demonstrates the wrong strategic approach to vehicle testing. With it, one cannot provide successful prediction of product efficiency. In reality, this article is only demonstrating how the usage of testing development process is declining, especially as one sees AV testing rising (Fig. 5.4).

Because of this unfortunate trend, the gap between the speed of development, design, and testing is increasing (Fig. 5.4), resulting in an increasing difference between design complexity and the testing accuracy.

Often a primary negative influence to effective testing is the goal of saving money for the testing process. Management seeks to reduce the costs associated with the testing, without taking into account the costs of poor quality of the product that are the direct result from the cheapened testing processes.

An analysis of the curve 2 demonstrated that:

- Testing advances have been very small.
- Those that do exist are primarily a result of increasing the use of modern system control in testing. Such advances are primarily related to the widespread implementation of electronics.
- By contrast, the methodology of accelerated testing is not improving, but in many cases actually decreasing, especially during the last dozen years.
- This trend is due to a declining emphasis on the real-world simulation that causes less investment in the testing technologies needed to accurately reflect the real-world conditions with production vehicle components.
- This trend stems from uniformed management desire to save the investment money needed to improve the quality of testing technology.

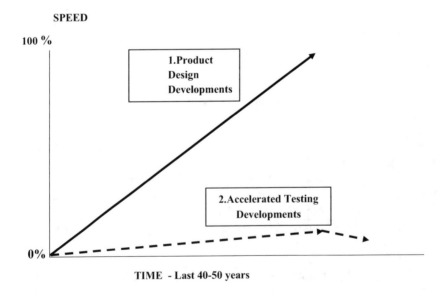

Fig. 5.4 Technical developments comparison

Figure 5.5 shows the trucks that Mack manufactured the past 100 years, and one can see how their design was dramatically changed for this time, stay much more and more complicated.

In 2005, the author visited the "Caterpillar 100 Years Exhibition" during an SAE World Congress in Detroit (Fig. 6.18). Displayed were models of their truck built during their century of manufacturing history. Figure 5.5 shows some of Mack's early trucks and one can see how their new models have dramatically changed from this time. Their trucks today have much more capability and, hence, are more complicated. Whereas an owner could once maintain his vehicle with simple tools, such is now impossible without advanced computer support. This means that early truck owners would accept lower reliability standards, because problems were easily corrected. Now, however, reliability standards must be higher to avoid expensive maintenance procedures involving costly repair tools. Achieving this reliability to avoid costs demands continuing improvements in testing technologies that the management has often not recognized. Unfortunately, the costs of these failures result in dissatisfied customers who must bear the burden of expensive repairs and deferred usage of the equipment that they purchased.

Unfortunately, when we examine the methodologies and equipment that companies use for testing these trucks, we will see little change over the years as was noted in the author's book *Accelerated Reliability and Durability Testing Technology*. He saw situation similar to Mack's testing program when he visited other companies such as Detroit Diesel. Clearly, the gap between the technical progress of their development technologies and their testing methodologies has increased during the years (Fig. 5.4). As a result, more problems with quality, reliability, safety,

Fig. 5.5 Mack trucks that were produced dozen years ago

durability, maintainability, and life cycle cost components of efficiency are being recorded in terms of vehicle recall incidents.

As has been mentioned several times in this book, too often the primary goal of testing development is saving project money by reducing testing costs. Paradoxically, the management does so at the risk of raising very expensive life cycle costs by not accounting for the degraded quality of their products. This is a serious and major policy failure.

We also see this philosophy at play with aerospace projects that are very expensive long-term projects. Here manufacturers are constantly seeking to save money by reducing testing costs. For example, when designing and manufacturing their 787 airplanes, Boeing had contracts with companies in 40 countries around the globe. In each case, testing by these contractors was not permitted. Specifically, some Israeli companies that designed and produced important components for Boeing did not have permission to test them. Failures occurred that resulted in lengthened development times from design to market. So, in an attempt to save testing dollars, the end result was greater expense and prolonged delivery time for the entire project.

For the past number of years, professional literature has documented significant advances in AV research, design, and testing. This echoes management's devotion of financial and engineering resources to this area of development. But it is concentrated on operational (driving) testing and not on reliability and durability testing. Moreover, there is a tactical selection of poor accelerated testing practices linked with the statistical methods being promoted in university laboratories that dismiss expensive and complicated equipment for physical simulation that correctly mirror real-world conditions.

Such statistical (theoretical) methods are easy to use in universities not needing the expensive and complicated equipment for the physical simulation that correctly mirror real-world conditions.

One clear indicator of these development trends is industry's usage of old, narrowly focused, and inaccurate approaches like the Arrhenius distribution that do not reflect real-world situation.

An example of such negative trends in testing can be found in an SAE webinar, sponsored by Element Materials Technology, in which was stated:

> As automotive battery technologies continue to advance, battery testing is a critical step to bringing a vehicle to market. Battery designs need to incorporate strong thermal management, shock and fatigue resistance, durability, fire resistance, and more while optimizing cost and offering long ranges and fast charging capability. With requirements increasing, battery testing needs to cover much more than cycling testing.

The above is true. The question then is, how does this company want to do it in practice?

> … an automotive testing expert will discuss the latest testing requirements, procedures, and best practices for battery cycle testing and beyond, including:
>
> • Multi-axis simulation testing, real-time durability/l=road load data acquisition (RLDA)

- Cooling systems, thermal management, pressure
- Enclosure testing, dust, spray, and chamber work
- Environmental testing: salt spray, humidity, and chamber work such as thermal shock, high temperature and humidity, and more

As we can see from the above comments, their batteries are being tested as a device separate from its other automotive components. This contradicts real life, where batteries do not alone, but as part of an entire vehicle, depend on it.

Correcting this anomaly requires a new vision about incorporating supply chain management concepts between the battery suppliers and the auto manufacturers. That is, the producers of batteries must partner with their automotive customers and develop batteries that meet the systematic needs of their vehicles. This change in management will signify a recognition that individual manufacturers can no longer see themselves as singular enterprises but must link together as a unified process that fulfills the satisfaction of their ultimate retail customers. This management philosophy implies that testing on an integrated, comprehensive basis must ensue as advocated in this and other books by this author. For more information about this concept, readers are encouraged to review *Supply Chain Management* by John Mentzer [9].

To analyze this situation further, because the webinar did not have a supply chain orientation, it could not show, for example, how thermal shock and high temperatures are higher than the maximum of real environments being endured by their customer's vehicles. Nor could the effects of these important differences be shown. It could not do so, because it portrayed the batteries as simply unique products that were being produced for a new electric vehicle market rather than vital components that had to be designed, tested, and produced for the demands of these new cars.

The producers of all components must adopt the same philosophy of supply chain management with their principle of production assembly manufacturers. The discussion of the batteries above is only one example of the critical efforts that must be made for modern testing methods to be incorporated fully into the development of new products.

The authors [9] wrote:

Definition: ALT is a method for stress testing of manufactured products that attempt to duplicate the normal wear and tear that would normally be experienced over the usable lifetime of the product in a shorter time period. If one has large number of components, component level test is required when they are underdeveloped.

This definition has a number of problems that are listed below:

- In the laboratory, it is difficult to simulate the real operating conditions on the machine's components.
- In real life, they interact with other components.
- Little value is placed on determining the absolute reliability.

Highly accelerated life testing (HALT) is a test in which stresses are applied to the product well beyond normal shipping, storage, and in-use levels. HALT has statistical difference with ALT.

Its disadvantages are the following:

- Not applicable in every case.
- Chemical degradation is not recognized as an important factor.
- Corrosion in a refrigerator door may not happen in a shorter time.

Acceleration factor (AF > 1) involves the following:

- Exposing test units to more than normal stresses.
- Higher temperature then highest temperature in real world
- Higher humidity
- Higher vibration then highest vibration in real world
- Accelerating chemical/physical essence of degradation higher than in the real world
- Weakened insulation of motor winding due to elevated temperature and moisture
- Weakened lubricant in bearing with exposure to moisture and high temperature

This article also said:

Appliance manufacturers have traditionally performed physical testing using prototypes to assess reliability and service integrity of new product designs. However, for white goods where service lives are measured in years or decades, the use of endurance testing to analyze long-time reliability is uneconomical. As accelerated life testing (ALT) is more efficient and less costly than traditional reliability testing, the methodology is finding increased usage by appliance manufacturers. In the present study, a simulation-based ALT approach was used to predict the service life of a polyacetal hinge which came from a consumer refrigerator. A predictive life stress model based on cumulative surface wear under accelerated stress conditions was developed and used to predict time to failure under consumer use. Results show that the life stress model demonstrated good agreement with performance testing data and reasonably predicts hinge life.

As we can see from the above:

- It was erroneously written that reliability testing is traditional approach. Really it is when one wrongly uses the definition for "vibration testing" or "proving ground testing" as reliability testing that it is a traditional approach.
- HALT and HASS mean simulation of two factors only—temperature and vibration—and sometimes +humidity, from many input influences.
- Both temperature and vibration simulation are higher than the highest temperatures and vibration level is recorded in real-world conditions.

One more common negative trend in accelerated testing—the separate simulation of field conditions—inputs for testing during design, manufacturing, and usage stages. Such separations do not take into account the real-world situation, where different influences (temperature, humidity, pollution, radiation, air fluctuations, field surface, human factors, and others) act simultaneously and in interaction with a product.

To be sure, there are various elements necessary for consideration in single groups, specifically the mechanical group, in the simulation and testing of field conditions for materials, details, and units in the real world. The various and different mechanical types of influences and testing act simultaneously in their interactions and not separately as is usually done in the laboratory testing.

Similar situations exist for other groups (multi-environmental, electrical [electronics], etc.) of conditions for accelerated testing.

Figure 4.5 in Chap. 4 depicts the complex interconnected influences that should be included in accelerated testing, but not commonly practiced for testing materials and vehicle components. Because this is so, they should be tested together and not separately as is usually done in laboratory testing.

The basic negative aspect of typically practiced accelerated testing is inaccurate simulation of field conditions. This is true not only just for an entire vehicle or components of machines but also in inaccurate simulation of these conditions for accelerated testing of simple units, details, and materials.

Figure 5.6 demonstrates the basic negative aspect of field conditions simulation—simulation of small part of input influences (such as temperature, humidity, and vibration) from many real-world input influences (compare with Fig. 3.3). Figure 5.6 depicts an *artificial construction of real world*, and not its natural conditions. The result of this artificial simulation is artificial testing, which leads to insufficient prediction of product efficiency (or reliability, or durability, or maintainability, and other efficiency components). But this negative approach is widely used, because it is cheap. This approach also relates to negative trends in development simulation, testing, and prediction.

Below is example of the many influences concerning corrosion testing shown in Fig. 5.7. As depicted, in the real world, corrosion is a result of the actions of two groups of influences—multi-environmental and mechanical. Listed below are those influences:

- Chemical pollution
- Mechanical pollution
- Moisture
- Temperature

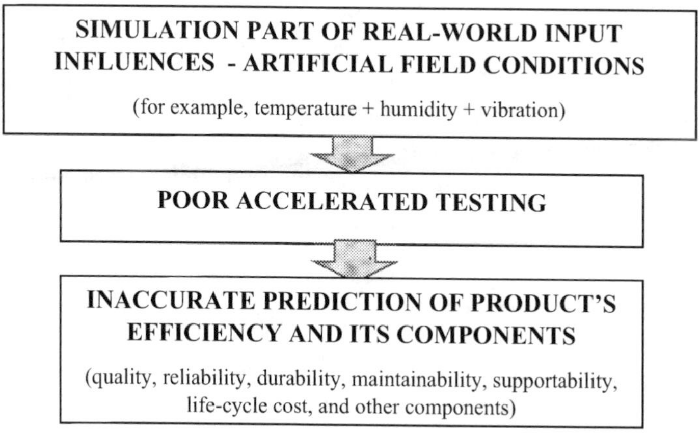

SIMULATION PART OF REAL-WORLD INPUT INFLUENCES - ARTIFICIAL FIELD CONDITIONS

(for example, temperature + humidity + vibration)

⬇

POOR ACCELERATED TESTING

⬇

INACCURATE PREDICTION OF PRODUCT'S EFFICIENCY AND ITS COMPONENTS

(quality, reliability, durability, maintainability, supportability, life-cycle cost, and other components)

Fig. 5.6 Path from simulation of artificial field conditions to inaccurate prediction of product's efficiency

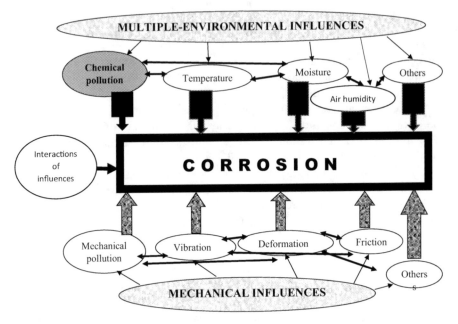

Fig. 5.7 Diagram showing how corrosion is a result of interacted multi-environmental and mechanical input influences and their interaction

- Vibration
- Air humidity
- Deformation
- Friction
- And others

But in practice corrosion testing, as performed by many companies and organizations, usually involves only one influence, chemical pollution or sometimes combined with humidity and/or temperature only. So, it is apparent that this laboratory corrosion testing does not include all real-world influences. Consequently, these simulations are not accurate, because they do not correspond to the corrosiveness of the real world. Altogether, corrosion protection system is based on the inaccurate testing results that may not be effective. This is a continuing negative trend in the development of corrosion testing.

A similar negative situation is with the discrete testing of units, such as engines or transmissions. Such practices do not simulate the vibration and environmental factors of the whole unit that is needed for accurate accelerated simulation and testing.

This type of testing is also typical for testing at the end of the design or manufacturing phases for the product. Of course, when this tactic is employed, some components could be subjected to more modern processes for the future development of these test subjects.

One more example is with the negative aspects of proving ground testing or laboratory vibration testing, wherein one practices to simulate mostly on a uniform surface profile. But in real-world vibration, especially mobile machinery, it is especially subjected to many more complicated processes as depicted in Fig. 5.8.

One can see from this figure that surface profile is one of many factors influencing a product's vibration environment. These are the reasons one needs to simulate all relevant influences, as shown in Fig. 5.8, for an accurate simulation of the real-world vibration effect on mobile machinery.

Another basic negative aspect of the trends in the development and use of modern systems of accelerated testing in many industries is the continued reliance on very old and narrowly established (e.g., old chemistry research data) and inaccurate statistical approaches such as the Arrhenius distribution and the generally used exponential distribution, which are not reflective of real-world situations. This negative aspect can be found in many industrial companies and organizations.

Figure 5.9 is an illustration of laboratory vibration testing equipment under old-fashioned negative methods.

Another example about negative aspects in development AT can be seen in the book [11]. This book contains 560 pages that are divided into 24 chapters and 177 subchapters. In this book, 560 pages text relative to testing is only general in nature and found in only 7 subchapters. Specifically, it describes the following:

• Subchapter, the role of testing (3 pages in Chap. 5)

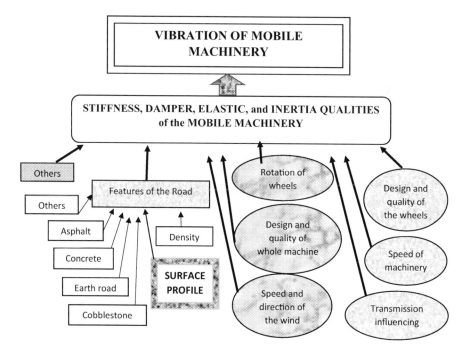

Fig. 5.8 Factors that influence the vibration of mobile machinery in the real world

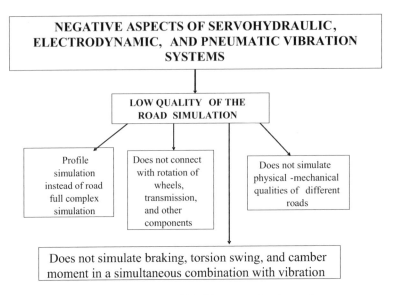

Fig. 5.9 Some negative aspects of current vehicle laboratory vibration testing

- Subchapter, product testing (2 pages in Chap. 8 "Reliability Growth and Testing")
- Accelerated tests (2 pages in Chap. 8)
- Tools and systems that support testing (5 pages in Subchapter 8)
- Equivalence testing (4 pages in Chap. 15)
- Some concepts and significance tests (1 page in Chap. 15)
- Accelerated life testing (4 pages in Chap. 22)

It is interesting to note that with this limited discussion of testing, the authors provided this description in their publication:

> The authors begin by presenting broad insights and high-level strategies for improving product quality. Next, they demonstrate how to implement robustness and reliability strategies that complement existing governance and decision processes. A section on tools and methods shows how to institutionalize best practices and apply them consistently. Finally, they tie strategies, decisions, and methods together through a case study project.

And "readers are introduced to many thought leaders whose writings can be sources of further learning."

From comments like these, it is easy to see why the trends of accelerated testing are moving so slowly or moving in the wrong direction.

Another example of negative aspects in accelerated testing development can be seen in [12]. Here the authors formulate the definition of accelerated life testing as a "method for stress testing manufactured products that attempts to duplicate the wear and tear that would normally be experienced over the usable lifetime of the product, but in a shorter period of time."

This definition includes the common formulation that relates to any testing, where one obtains test results in a period that is shorter than that experienced in normal usage. But this definition of accelerated testing is too general and does not provide any specific requirements. It really does little to advance the art and science of testing.

The authors actually confirmed this when they wrote:

Problems

- It is difficult to simulate with operating conditions
- In real life, they interact with other components
- Little value in determining the absolute reliability

The authors then formulate their definition for highly accelerated life testing (HALT) as:

- A test in which stresses are applied to the product well beyond those experienced in normal shipping, storage, and in-use levels.
- HALT is scientific.
- HALT has statistical differences with ALT.

But they offer little evidence as to what is meant by scientific or what specific statistical differences exist between HALT and HASS.

These authors then write further:

Accelerated factor AF > 1:

- Exposing tests units to more than normal stresses
- Higher temperature
- Higher humidity
- Higher vibration

Accelerated chemical/physical degradation

But they remain mute on other real-world factors that influence product degradation and failures.

These are a few of the examples of how some current publications are not embracing the advances needed in accelerated testing.

Excerpt from the seminar titled, "Autonomous Vehicle Test and Development Symposium" [13]:

Engineers and managers involved with product development are constantly challenged to reduce time to market, minimize warranty costs, and increase product quality. With less and less time for testing, the need for effective accelerated test procedures has never been greater. This course covers the benefits, limitations, processes, and applications of several proven accelerated test methods including accelerated reliability, step stress, FSLT (Full System Life Test), FMVT® (Failure Mode Verification Testing), HALT (Highly Accelerated Life Testing), and HASS (Highly Accelerated Stress Screening).

A combination of hands-on exercises, team activities, discussion, and lecture are used throughout the course. Participants will also receive a copy of the instructor's book, *Accelerated Testing and Validation Management*, which includes numerous hands-on exercises and a CD with analytical spreadsheets.

This symposium's program offers yet another example of the poor attention given to the role of and importance of testing. Among its many presentations, only three were related to testing. All of them were related to proving ground testing that, as will be discussed later, can evaluate a product only under proving ground conditions and not the real-world conditions that alone can provide accurate predictions for such components as reliability, durability, safety, and maintainability. This is because proving ground testing alone cannot account for the totality of real-life environmental influences, including customer usage changes during the product's service life.

During the early 1990s, the European nations conducted an exhaustive research program about the feasibility of autonomous vehicles. Entitled the Eureka Prometheus Project in Paris, it acknowledged how self-driving vehicles were becoming a reality. Yet 20 years later, the final stages of their testing, validation, and fail-safe testing efforts still pose a huge challenge to the automotive industry. The rigorousness and thoroughness of their testing processes need to be at a level of accuracy and effectiveness than anything that has gone before if the final goal of safe, effective autonomous vehicular travel is to be achieved [14].

The Autonomous Vehicle Test and Development Symposium is one of Europe's largest autonomous vehicle events [14].

Here can be seen an example of both superior testing methods and of poorer testing methods. First included in their report a better testing method was the following statement, "Accelerated durability testing of automotive components has become a major interest for the ground vehicle industries."

The authors of this comment believe that this approach can predict the life characteristics of a vehicle by testing fatigue failure at higher stress levels within a shorter period of time. Current laboratory testing generally uses rigid clamps to mount the component to a shaker table. The authors correctly acknowledged how this approach is not accurate for durability testing of most vehicle components, especially those connected directly to the tire and suspension system. But, for their purposes in this experiment, they studied through experimental testing and numerous simulations the effects of elastic support on their tested structures. This method effectively demonstrated the effects of natural frequencies, damping ratios, and mode shapes, so as to provide more accurate estimates of structural fatigue life spans.

Let us now examine why their techniques were superior. First, a specially designed subscaled experimental testing bed with both rigid and elastic supports was developed to study the effects of the additional elastic support and the mass on the change of structural model parameters. The elastic support mounts recorded significantly wider variations for the model parameters than the rigid mounts. Next, the model parameters with elastic support were then used [14] to build and tune the finite element model (FEM) [14].

Now, an example of a poorer testing method is being used for ground vehicle testing [15]. This study used a century-old proving ground vehicle testing concept for evaluating the durability of modern product in the real world. These authors wrote in [15] that such vehicles should be subjected to rigorous testing in the field and on proving ground tracks and that the vehicle's response to each be recorded. The acquired road load data would be analyzed and pseudo damage values estimated from both field- and proving ground data.

Using this methodology, which is often in current practice, proving ground sequences are derived from correlations of the damage produced in customer usage conditions and from accelerated proving ground tracks data. These correlations required the structural frequency responses gathered from the field to be incorporated in the proving ground testing data with their combination providing final product durability validation estimates.

The authors claimed an improvement of the quality of "accelerated proving ground (PG) durability testing" was derived by frequency-based pseudo damage spectrum (RDS) method.

A final use profile and intendent performance profile was developed based on testing surveys and was finalized as presented in [15]. The actual customer routes that were used in the field trials were selected subjectively to represent usage profile [16]. These authors wrote that the next step would require road load data development in both the field- and proving ground conditions. For proving data measurements, trials were to be undertaken a different speed at each track to capture a wide range of frequency responses.

The next important negative aspect of this approach was the development of a correlation methodology for comparing proving ground testing results and field results. The following phases were utilized to establish an "optimized" proving ground durability test sequence:

- Terrain classification
- Mixed calculation and target setting
- Power spectrum density (PSD) analysis
- Frequency-based damage correlation

For the last bulleted item—frequency-based damage correlation—the authors calculated the potential damage incurred from both proving ground and field road load data by using a common artificial summation notation (SN) curve and slope, as was published in [16].

In [17], the authors then concluded that "this optimized proving ground durability test sequence can be developed for each commercial vehicle model by developing frequency-based correlation of road load inputs of customer usage. These test cycles give a competitive edge to vehicle manufacturers over others for providing the customers with a more durable product and for incurring less developmental and warranty costs."

But, as was stated earlier by this author, proving ground testing cannot duplicate or simulate the plethora of multi-environmental factors found in the real world. This being the case, it seems obvious that the testing engineers simply ignored them and conducted their tests with the few variable they could simulate. Therefore, proving ground testing alone is not an adequate approach for durability reduction or degradation estimates for field operations. Any proposed methods of recalculating proving ground test results as a substitute for actual field test results cannot represent the physical essence of real-world actions. But as proving ground testing is less expensive and simpler than serious durability (using reliability and other test methods) testing, this negative trend in the development of accelerated testing continues.

The example below relates to an erroneous use of vibration testing while calling it reliability testing [18]. The 14th Institute of the China Electronics Technology Group Corporation is the birthplace of China's radar industry. It has developed many pieces of high-end radar equipment and its products have been exported to dozens of countries and regions.

The institute has been using the internationally accepted "m = p" VibControl vibration control systems since 1998 to conduct sophisticated radar reliability tests. In this article was written:

> The vibration tests carried on shakers and controlled by "m + p" VibControl systems therefore play a key role in reliability verification. The 14th Institute has more than 10 sets of "m + p" VibControl Software with m + p VibPilot and m + p VibRunner acquisition hardware, m + p VibPilot is a compact, rugged 4/8-channel hardware platform. For higher channel counts, m + p VibRunner hardware is the first choice. It can be used as a desktop instrument or mounted into a 19in rack and supports distributed measurements. Equipped with 24-bit sigma-delta A/D converters and a sampling rate up to 204.8 KHz, the m + p acquisition hardware allows for alias-protected measurement in a frequency range up to 80 kHz and with more than 120 dB spurious-free dynamic range.
>
> Following the strategy of always striving to be a leader in the aerospace and defense testing sector, m + p international has integrated advanced control capabilities such as nothing/force limiting into its VibControl software. Many safety features ensure reliable closed-loop vibration control—from pretest checks to abort checking, nothing, and controlled shutdown. The radar's reliability level is constantly increased using these tests. Finally, the expected value of mean time between failures is reached and its operating cost is reduced. In many radar reliability tests, m + p international's products played an important role. The responsible technical engineers of the 14th Institute expressed their satisfaction with the test results and gave a high evaluation of m + p international's products [18].

As we can see, this institute uses vibration testing, which is one of many components of reliability testing, but calls it "reliability testing." This simulation incorrectly separates influences instead of a complete complex of simulation, as was described earlier in this book, as well as others of the author's books discussed real-life reliability testing.

Let us consider one more work from the Automotive Testing Expo 2021 [19].

A specialist in advanced testing technologies for active vehicle safety in 4active-Systems wrote how they:

> … provide solutions to reduce road fatalities and ensure safety and compliance with the highest international standards around the globe.
>
> The company's latest product innovation is its highly automated robotic platform 4activeFB-eco for future ADAS and autonomous driving (AD) testing. It is said to be the flattest robotic carrier for VRUs like pedestrians or bicyclists, with the lowest radar cross section of all existing carriers today. It is designed according to Euro NCAP VRU and related ISO standards. Open standards (OpenDrive, OpenScenario, etc.) are used as input to create scenarios from external resources like simulation tools, real road testing, software/hardware-in-the-loop tests or existing databases from, for example, accidentology and standards regulation. An Open Control Center according to the upcoming ISO-WD-22133-1 is applied to control and monitor all devices and infrastructure at the proving ground. All tools communicate via a dedicated Wi-Fi mesh system, so complete coverage of every existing test track is guaranteed. Automated reporting tools significantly reduce the time for post-processing test data. 4activeSystems' wide range of established and certified dummy objects (cars, pedestrians, motorcyclists, etc.) and infrastructure (light systems, obstruction

walls, etc.), as well as future and customized targets (animals, roadside furniture, etc.), complete the technology platform. This novel closed toolchain enables seamless testing and completely covers increased demands for future AD testing on proving grounds.

The above solution relates again to the fallacy of exclusively using proving ground testing, which does not simulate accurately the real world as described in [1–3]. Therefore, the above solution could not be useful for generating successful prediction of product reliability, durability, maintainability, and other components of efficiency.

5.3 Trends in Using Virtual (Computer) Simulation and Testing as a Replacement for Real-World Conditions

Virtual testing is the future of modern testing technologies, but it is currently inadequately developed, because it is not based on accurate simulation of real-world conditions. Still it is popular, especially in academia, because it is cheap by eliminating the need for large laboratories or the expensive equipment **for** physical simulation technologies.

Many articles, books, and papers have been published lauding how virtual simulation solve engineering problems. But many professionals often use this approach without considering carefully whether virtual simulation truly provides accurate reflections of real-world vehicles and their usage conditions.

There are many examples of this problem and let us now consider a few.

According to Capes Jones as stated in IT World [10]: "Poor software quality costs $150+ billion per year in the USA and over $500 billion worldwide." Many of these software quality issues come from poor test data quality.

And:

According to NIST the average testing team spends between 30 and 50 percent of their time setting up test environments rather than on actual testing, and the estimated number of projects with significant delays or quality issues is 74%.

Fortunately, one solution to the test data quality problem is through the use of a technology called virtual data. Similar to the way that virtual machines create virtual copies of physical computing resources, virtual data creates multiple lightweight virtual data copies from a single, full-size copy [19].

The value of virtual data can be seen in [10]:

The problem with test data is that fully testing if code is ready for production requires a parity copy of production data, yet creating full parity copies of production data is often too onerous for most QA teams to manage. In order to more easily manage test data in development and QA, teams often use subsets of production data. Then before releasing final code to production, the code is run on a full-size copy of production data. This final testing might be done in the last weeks of a multi-month project. What typically happens is that this final production parity testing flushes out more bugs than can be fixed before the release date. Thus, the release has to either be delayed or released with bugs.

One common negative aspect of all this is an increasing usage of virtual testing and a diminished usage of physical testing. This situation can be seen in Fig. 5.10.

The author has written about this in his previous publications, including [1–3], and others. The basic substance contained in these publications is as follows:

- Virtual testing is easier and less expensive.
- The algorithms presently employed in the software programming of test subject cannot yet amalgamate the varied interactions present in complicated machinery. Generally, they separate their simulations to discrete components or units that are not accurate representations of real-world interactions. Some researchers sometimes try using mathematical analysis, such as the Monte Carlo distribution model, but even these cannot accurately simulate the real interactions of a product's components (details and units), especially in complicated machinery.
- Algorithms need to be developed for software programming that better simulate the real-world conditions such as the combined effect of solar radiation, real components surface variations and fluctuations, and many others.
- While virtual simulation of real-world conditions can be a very useful tool in the predesign process, companies are increasingly relying on virtual simulation for the design and the manufacturing phases where their operating environments are much more complex and mirroring them through simulation is more difficult.
- This increase in the use of less expensive virtual accelerated testing also encourages decreased investment and development of the more difficult and expensive physical simulation of field conditions and the investments necessary for the

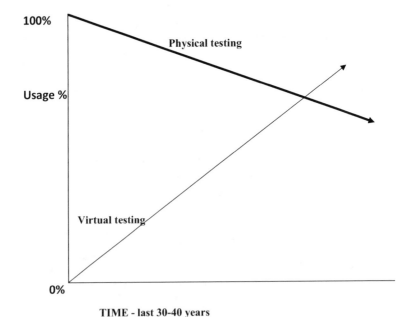

Fig. 5.10 Comparison of testing approaches usage

many types of accelerated testing. This can be seen in Fig. 5.10 that compares Expo's accelerated testing methods and equipment of 20–30 years ago to that presently available.

These observations can also be seen in the publications of other authors. Although many of them believe that virtual testing provides major benefits, primarily in economic and time-to-market savings, they neglect to account for the disadvantages inherent in virtual testing as compared to physical testing. Specifically, this includes the use of inaccurate simulation of real-world conditions and their interaction(s) with the real-world environment and interactions between different components of the product.

Physical testing has evolved from a long history of developed processes aimed primarily in the quantity area and much less on the quality area. Some of this testing accounts for the complexity of new products and their many components interacting under real-world conditions (see [1, 2, 4]. But seldom do these processes rise to a level appropriate for true accelerated reliability and durability testing technology (ART/ADT) that is based on accurate physical simulation of real-world conditions. Virtual testing can rarely take these complexities into account. Consequently, it is not presently capable of providing accurate simulation that lead to successful prediction, which then leads to increased recalls and increased costs as has been detailed in this book.

Consider this example that John Wilson wrote in his article "Why So Many Product Recalls" [6]. He reported that a government website https://www.recalls.gov

> ... provides recalls information lists more than 60 automobile recalls, plus approximately 24 recalls for tires in 2017. The recalls involved a wide range of manufacturers and vehicle types. In addition to the automotive-related recalls, approximately 300 consumer product recalls were issued in 2017.

He wrote further that military and aerospace products, which generally are not included in consumer recall data, also continue to have problems as new high-tech designs are put into operation. New, improved launch vehicles, missiles, aircraft, weapons systems, and ground vehicles are still experiencing excessive field failures. Then John Wilson asked: "...what has happened to quality design and manufacturing? Why do we see so many recalls?"

Here is his answer [6]:

> ...several factors came to mind. There seem to be more new products introduced than in the past. Many products are more complex than in "the good old days," and more complexity means more things to go wrong. Just look under the hood of the late model car.
>
> Also, today's products include more "high-tech" features that may not be as tried and true as they should be before being put to the market. Everything from dolls to light bulbs and appliances plus all of the "Internet of Things" gadgets have problems. Another factor is more diligent monitoring and reporting.
>
> Twenty or 30 years ago, there were fewer product safety monitoring and reporting agencies, so consumers were just stuck with faulty (and sometimes unsafe) products. Nobody knew and nobody kept track and reported.

In this reference he also stated that:

…while these are all contributing factors, it is also true that we also have more sophisticated design and quality assurance tools. The same technology that lead to the introduction of modern products can also be used to improve these products and to reduce cost. For example, thanks to integrated circuits and digital electronics products that use far fewer discrete components to accomplish the same functions as 20 or 30 years ago. Theoretically, this should produce greater reliability due to fewer components, fewer and sturdier connectors, and greater understanding of the failure mechanisms.

He further wrote:

… if technology is such a benefit, why do we still have so many inadequate or unsafe products? As an old graybeard engineer in the testing business, I have a tendency to blame all of the fancy, high-tech electronics. However, I think a thorough statistical analysis will prove that false. Many, if not most, problems are, in fact, mechanical. Some component breaks, binds, slips, wobbles, or wears prematurely. So, what is the cause of most of these problems? Why do these components not functioning as intended, as the computer models suggested?

Wilson finally postulates further that too much physical product testing has been eliminated from the product development process. This is largely, because it is so much faster, easier, and less expensive to model and analyze problems on computer. Unfortunately, virtual testing is only as thorough as their programs, algorithms, and assumptions, and they are not able to foresee or account for all of the nearly infinite variety of conditions misuses, abuses, and manufacturing variables that products experience in real-world operation.

However, Wilson did not propose doing away with computerized design and analysis or even reducing the increased rate of their use. Rather, he proposed that physical testing should be reinstated and even improved and that the feedback from such efforts can upgrade existing computer models. Unfortunately, it is also true that regardless of any progress made in simulation and testing, there will always remain the prospect that until the products are actually misused and abused, we will continue to see more recalls.

Another example comes from Joachim Linday, the Mercedes-Benz' senior manager, of Overall Testing for E-Class cars who said in [7]:

We can computer simulate each under-hood component and the temperature in the area in which it operates and how cooling air can be introduced. For example, we can do that for the wiring harness. But all that is design: to be certain that everything works, it still has to be physically tested because in the computer you do not have a precise picture of the real world.

Then he said:

I believe that we always need to do final testing physically, because the customer doesn't buy an electronic program, he buys a car!

There are several driving forces that are reducing field or physical product testing and replacing them with computer simulation. These include as follows:

1. Publications regularly emphasizing any trends in engineering development that are based on the product's development that is accomplished through computer simulation and accelerated testing.

   For example, James Truchand, president and CEO, of National Instrument wrote [20] that

   > ...for most of the last 100 years, a special focus has been placed on quality and testing that was traditionally physical testing. However, this has been changing over the past decade. This is partially due to significant emphasis in reducing automotive product cycles from the previous norm of 4–5 years to present requirements for 12–18 months. This change called "zero prototyping" was only recently introduced. As a result of this trend, simulation needed to be added formally into the design cycle, resulting in a "simulate-build-test-ship" design cycle. The dependence on simulation early in the design cycle intensified the need to institute simulation by mathematical models with test result much earlier in the product development cycle.

2. Another recent trend is the increasing complexity of products being coupled with the replacement of mechanical systems or structures through the use of electronics diagnostic tools. As the vehicles' electronic content increases, most of the testing, which was previously primarily mechanical in nature, now incorporates the added complexity of electronics testing. This presents the requirement for a common test platform that addresses seamlessly both the mechanical and electronic test environments.

   This increased complexity also means product designers and manufacturers have had to confront a multitude of challenges, as product specifications have become more complex. One example of this is the addition of tire pressure monitoring systems to vehicles. Also due to globalization, a typical vehicle platform which was originally designed only for the US market must be reengineered for European and Far Eastern markets. It is shared across assorted brands worldwide and ultimately becomes the basis for vehicles in several classes manufactured in different plants around the world and complying with a multitude of regulatory requirements.

3. The integration of physical testing with virtual testing. In the product development process, testing appears in two areas: first upstream as a means for establishing product designs and performance requirements, and later downstream as a pass/fail criterion on the product prior to its release for production. Although the types of measurements in both instances are similar, the tests' purposes how an organization uses the results of the testing are substantially different. This means that prior to physical testing, virtual simulation can help engineers identify the optimal predesigns or challenges that exist with the prototype and should be limited to that purpose.

A bidirectional flow of information between the simulation used in design and the information gained from product performance testing is critical for success. Traditionally engineers compared testing data from previous models or components against simulation results and used the information to calibrate new testing data, thereby increasing the confidence in their simulation predictions for the current designs. Test data from previous models and components can also be used as inputs for new simulations to improve the fidelity of their results. Simulation can also provide insights that permit minimizing and optimizing testing. Determining the optimum location of sensors, actuators, and exciters is one such example.

Another important aspect of this integration is the need for an early test platform to deliver connectivity to the design and simulation tools. Simulation software vendors need to build better connection with test platforms moving forward as has recently happened over the past decade with computer-aided design (CAD). This should also include addressing the growing need to provide integrated methods that allow better visualization of, and comparison of test and simulation data, especially through the increased use of video. But paramount is the need for virtual simulation and integration testing to reflect real-world processes, as has been detailed earlier.

4. Included in these multi-domain and multidiscipline nature of modern engineering system design and testing is another trend that greatly affects testing methodologies. Consider, for example, the automotive entertainment systems now in popular use.

This once simple and discrete component, a radio, has been completely transformed into an automotive information and media center. Today's "radio" design and testing now must include assuring the performance of TV display(s), MP Player, DVD and CD players, FM and satellite radio, GPS navigation systems, cell phone and e-mail access, electronic games, remote diagnostics, satellite-based car alarm and control, and other features. This is a major change from the original AM radio function.

Similar challenges exist with other systems such as power train management, climate control, intelligent braking and handling systems, and other features. The quality assurance and fail-safe design of such complex systems requires, first, a dramatic development and use of ART/ADT, multi-domain measurements, and excitation capabilities in a modular and expandable test platform that must be executed in a coordinated and time-critical manner.

5. The need for a next generation of physical test platforms. With the significant changes occurring in engineering accelerated testing methodologies, it is unlikely that prior fragmented approaches to testing will be able to scale their capabilities appropriately. There is a growing need for an easy-to-use robust modular, customizable commercial-off-the-shelf (COTS) test hardware and software testing platforms with plug-and-play architecture that is akin to CAD. It must be robust enough to alleviate hardware connectivity issues to make the hardware transparently intuitive. It should also provide seamless connectivity to the environment surrounding the test through the use of communication buses such as CAN (Controller Area Network), sensors and actuators with Transducer Electronic Data Sheet (TED) capabilities, simulation Product Life Cycle Management (PLM) software, Enterprise Resource Planning (EKP), and manufacturing execution systems. The National Instruments (NI) LabVIEW [21]

graphical test software platform is one such example that provides the capabilities cited above.

LabVIEW integrates seamlessly with hardware and provides built-in connectivity to third-party devices and the test environment.

As was discussed earlier, the basic weakness of the above trends in testing development is their inaccurate simulation of complicated product technologies and their interactive components. Therefore, virtual simulation can now be useful mostly only in predesign processes. If these tools can be improved enough to meet the demands cited above, then their productive usage will become more effective and less expensive. The author's opinion is simple and must be repeated here. The most effective technology is that which combines physical testing with virtual testing.

5.4 Consideration of the Newly Trends in the Development of Real-World Simulation, Accelerated Testing, and Efficiency Prediction

One ongoing need is for standardized terminology and its consistently proper usage by everyone involved in the simulation and testing profession. A good example of sloppy issue can be seen where an author [22] used misleading terms for vibration testing and environmental testing. Specifically, he stated:

> Environmental dynamic testing is a technical discipline that includes all vibration tests conducted on most engineering structures. The goal is to simulate the effects of the operational environment on a given object. A car clutch, a dishwasher pump, or an airplane altimeter are just a few objects required to pass a dynamic environmental test prior to use.

Then he continued:

> In general, there are three main stages in vibration tests: first is test setup. This phase is critical to test success and the actual component life span relies on a good test setup. Two tasks need to be accomplished: define a test profile that represents the operational vibratory environment and fix the test item to the shaker in a way which represents its real operational mounting. In many cases test profiles are taken from standards.

This is problematic from several aspects. First, "vibratory environment" is a combination of two different influences—vibration and environment—that must be applied in a combined fashion. Second and very troubling, to duplicate real-world conditions, the test conditions must reflect the real-world situation, not an artificial profile comprised of arbitrary standards.

Third, Latanne wrote [22]:

> The second stage is the test itself. During the test, the vibration controller is the 'main player.' This statement generates lots of questions. Single or multiple inputs? Response limiting? How many statistical degrees of freedom (random test) or which compression factor (sine test)?

Qualified professional know that vibration in the real world has a random character.

The real-world inputs are multiple, not singular, and that for a twenty-first century testing protocol, the real vibration of mobile equipment must have six degrees of freedom [3].

As can be seen from this article that was contained in a respected international journal, its failure to use technical terms properly has created a conflict within its title "Back to Environmental Basics" which implies "go back to the twentieth century" [22]. Such was not the author's intention, but his poor word usage created this illusion.

To demonstrate further how the progress of accelerated reliability and durability testing, as well as successful prediction, is moving forward very slowly, let us review the following current magazines that specialized in automotive and aerospace testing areas.

There is an edition of a dedicated magazine called *Automotive Testing Technology International* that was published in November 2020 [24].

It is one author, Atsmon, CEO, Cognata (Israel), authored an article "Data Farming" and stated:

> When it comes to autonomous vehicles, there is no margin for error. To absolutely guarantee safety and user confidence the AV needs to detect everything from physical objects to micro-discrepancies in visibility levels. Even variables on the car itself such as tire and brake wear need to be anticipated. The purest way of getting an AV to correctly react to these elements is to place it on the road. But even after millions of kilometers, real-world trials can only do so much: virtual testing is required to complete the picture.

Then Atsmon wrote:

> Simulation does have its own limitations but these are gradually being addressed. One problem being solved right now concerns data annotation. Typically, vehicle developers would collect data sets from driving scenarios in the real world before applying these to a simulation program. To apply the data onto objects that are appearing in the virtual environment, such as redestrians and cars, they would need a team of people to manually annotate everything one frame of footage at a time. However, new simulation approaches are set to relieve this task and give developers more time to refine their perception products, from cars to cameras to cheap. British simulation specialist rFpro producers digital-twin models in which users can conduct tests. With its new method of simulation-based automatic object annotation, rFpro claims it can slash the costs and error risk associated with the standard manual technique and open AV technology development to more participants.

Atsmon then quoted Mr. Matt Daley, managing director of rFpro: "The key switch in the past few years has been that we don't need to constrain ourselves to real time."

And finally:

> We can utilize this data throughout the simulation, so it saves lots of manual annotation and lots of time, plus you get higher-quality data.

Atsmon went to explain:

> "We use deep neural networks to learn how sensor technologies react in the real world."
> "We are then using the networks to transfer the synthetically generated data from our engine

into something that looks like the actual sensor. It's called transfer learning. You can utilize this data throughout, because this process is fully automated."

As we can see from the above article:

- It was considered simulation for technological processes, which means the quality of driving processes, and nothing about simulation for reliability/durability (ART/ADT) testing.
- Therefore, this simulation and testing cannot offer accurate initial information for successful product's efficiency prediction.
- This demonstrated the narrow thinking and critical thinking on the part of these companies.
- This and other articles in the magazine ignored physical testing and considered only digital testing that cannot be useful enough for industrial companies wanting real-world information for the successful development of their product.
- Atsmon's incorrect usage of "real world," when, in fact, this was an artificial world.
- Finally, Atsmon and Daley do not take into account that driving is only one of many components of the real world.

In total, *Automotive Testing International* which published Atsmon's article consists of 63 other articles, of which only 20 directly involved testing. Even then, among them, ten focused on proving ground testing that ignored multi-environmental testing. The remaining articles and interviews were about functional or operational testing. As was considered they do not simulate accurately the real-world conditions. For example, this type of testing does not simulate the corrosion processes during a machine's life cycle nor how human interventions (operator's and management) influence the product's reliability, durability, life cycle cost, and many other efficiency components. Therefore, it cannot even be used for accurately not only for efficiency prediction, let alone for reliability, durability, and other efficiency component successful prediction. As we know, accurate prediction is the final goal of testing.

By contrast, *Automotive Testing International* recorded compressed development time and reduced testing cost as industry's top priorities, particularly as vehicle makers transition into an electrified future. Waiting on the environment, climate, weather, terrain, and the logistics of on-road testing were seen merely as costly impediments, when the program clock for market release was ticking. For commercial vehicle manufacturers, the need for comprehensive testing facilities that offer complete seasonal independence and utmost process efficiency has been growing. Unfortunately, their drive for such facilities is focused on operational issues and not reliability and durability issues.

Atsmon also wrote in [25]:

When it comes to autonomous vehicles, there is no margin for error. To absolutely guarantee safety and user confidence the AV needs to detect everything from physical objects to micro-discrepancies in visibility levels. Even variables on the car itself such as tire and brake wear need to be anticipated. The purest way of getting an AV to correctly react to these elements is to place it on the road. But even after millions of kilometers, real-world trials can only do so much: virtual testing is required to complete the picture.

He went on to write:

Simulation does have its own limitations, but these are gradually being addressed. One problem being solved right now concerns data annotation. Typically, vehicle developers would collect data sets from driving scenarios in the real world before applying these to a simulation program. To apply the data onto objects that are appearing in the virtual environment, such as pedestrians and cars, they would need a team of people to manually annotate everything one frame of footage at a time. However, new simulation approaches are set to relieve this task and give developers more time to refine their perception products, from cars to cameras to cheap. British simulation specialist rFpro producers digital-twin models in which users can conduct tests. With its new method of simulation-based automatic object annotation, rFpro claims it can slash the costs and error risk associated with the standard manual technique and open AV technology development to more participants [24].

Matt Daley, managing director of rFpro was quoted in Atsmon's article as saying [25]:

The key switch in the past few years has been that we don't need to constrain ourselves to real time.

Continuing on Atsmon explained:

We can utilize this data throughout the simulation, so it saves lots of manual annotation and lots of time, plus you get higher-quality data. We use deep neural networks to learn how sensor technologies react in the real world. We are then using the networks to transfer the synthetically generated data from our engine into something that looks like the actual sensor. It's called transfer learning. You can utilize this data throughout, because this process is fully automated [25].

The reader can see how this meaning corresponds to the issues raised in this chapter about the negative trends arising in testing development. As we can see from the above paper:

- It only considered simulation for technological process testing, that means the quality of autonomous driving without human control and nothing about simulation for reliability or durability testing.
- Therefore, this simulation and subsequent testing cannot offer accurate initial information for successful product's efficiency prediction. Specifically, automated driver systems are being tested to respond only to external conditions such as traffic, signal lights, and pedestrians. Such testing ensures that cars will slow down, change directions, or stop as needed. However, when an internal emergency arises due to product failure and the human driver has been lulled into complacency or poor driving skills, the automated system will not be able to react safety. A catastrophe will ensue. Thus, it is critically important for automated cars to be designed and tested exhaustively under real-world conditions to ensure reliability and product safety as has been discussed in depth in this book.
- This reflects the narrow thinking of this company professionals, as opposed to the deep thinking they claim to utilize.

- This and other articles of the magazine have considered only digital testing and ignoring physical testing that cannot be useful enough for industrial companies making real-world prediction.
- The article's authors use the term "real world" in these articles incorrectly when in fact there is an artificial world of assumed situations.
- Finally, the authors do not take into account that driving is only one of many components of the real world.

In the [26] it was written that:

...Proventia, a Finnish technology company that specializes in modular test facility solutions, offers a new range of test facilities with efficiency and flexible features for today's fast-moving automotive industry. Proventia's test solutions are suitable for the development and testing of electric and hybrid vehicles and their power trains, battery packs, and hydrogen fuel cells.

Modularity is the key feature in Proventia approach to test facilities.

Proventia test cells are compact modules built specifically for testing purposes. They are structurally and environmentally designed to meet the high performance standards of a testing facility.

And:

Proventia modules are spacious cell with insulated walls and a high structural integrity. A small test rig is often just one module, including all the necessary components. Larger test facilities for more complex testing procedures such as multi-axis testing and full vehicle testing can be cleverly constructed by combining multiple modules with rooftop structures for additional supportive equipment.

The above article, such as shown in Fig. 5.10 above, does not include any new methodology, because:

- For example, this author described in his book *Trends in Development of Accelerated Testing for Automotive and Aerospace Engineering* [23] how the modular test centers improve a higher quality of simulation.
- Earlier (in 2016) SAE International published the author's book *Successful Prediction of Product Performance*, where its Fig. 5.18 illustrated the essence of modular test center for whole vehicles or their units. Therefore, the technology in considered paper is not new.
- Moreover, the author's above-referenced publications demonstrated how higher level of test centers make possible accurate accelerated reliability and durability testing. However, the magazine articles cited above include only primitive testing technologies such as multi-axis vibration testing without combining them with other types of testing.
- Finally, the above article is demonstrating known earlier trends in accelerated testing development, but not an innovative quality of testing.

Let us consider another journal *Showcase*, which is published by *Aerospace Testing International* [27]:

It includes 40+ papers mostly related to flight testing or small units testing. Among its articles were "Flight Test Roundup," "Flying in the Tech Center's Chopped

Up Chopper," "The Proper Test for High-Quality Lamps," "Advanced Materials for a New Aerospace Reality," "Selecting Cameras for High-Speed Imaging Techniques," "Assessing Additive Material Safety for Aerospace Industries," "In-Flight Pressure Measurement," "World's Smallest Lepe Triaxial Accelerometer," "Flight Test Instrumentation System Design and Integration," etc.

Many of them discussed vibration testing, where there were a few new details of this type of testing. But they did not consider the development of accelerated vibration testing essential technology, because they consider only one input influence— surface profile—instead of many real-world influences on the product's vibration profile demonstrated in Fig. 5.7 above and analyzed in detail by the author in his books [1, 3, 4].

Some of these referenced papers related to essential development of new testing system. For example, in his paper *How to Future Proof Your Test System* [27A], Dave Baker wrote: A system components age and replacement parts become harder to find and more expensive to purchase, and unsupported systems are more vulnerable to breakdowns and cyber security threats.

All too often, addressing diminished support and scarce part availability caused by obsolescence is an afterthought when problems arise. These obsolescence problems are typically further magnified because many organizations are operating monolithic systems using custom hardware and propriety software. However, there are ways to design systems that take a proactive approach to managing obsolescence, as well as methods for maintaining existing outdated systems that make managing aging aging systems far less expensive and time consuming [26].

And:

Like speaking brakes, test and measurement systems develop warning signs when they are on the verge of failing and potential victims of obsolescence. Two major indicators that a system is approaching the need for modernization include slower execution and more frequent failures and errors. Ignoring these two problems will eventually lead to equipment breakdowns that can leave you stranded as prolonged downtime is usually needed to retrofit solutions to existing components. Oftentimes, these changes need to be made with limited support from vendors for the nature equipment.

Another warning sign that your system may be on the verge of needing an upgrade is when compatibility issues prevent the addition of new features into your system. For example, an old operating system (OS) may not support the addition of new hardware that could accelerate your productivity. Similarly, aging systems may also experience security breaches or cyber events, because of an older OS or the fact that the system is running on an unsupported platform. In short, for a managerial point of view, all these signs of the need for modernization come back to the fact that the system is running slower and shipping away at productivity. As a result, you are spending more money on maintenance and productivity decreases [26].

As we can read from the above article, the proposed testing of technological (driving testing) process is not connected to reliability or durability testing, evaluation, or prediction. Therefore, it is not useful for either the successful prediction of components or complete product efficiency.

The goal of the paper looks like testing for testing sake alone.

Let us consider the next paper with an ambitious title "Developing a National Satellite Testing Capability" [28]. The authors, Macintosh and Belly, write that "A major new national satellite testing facility in the UK is to receive vibration testing equipment from one of the aerospace sector's leading suppliers."

As has been repeated many times in this book and other publications by the author, vibration testing is only one from many types of testing needed for providing product's efficiency evaluation or prediction. Therefore, a satellite vibration testing facility cannot be a major new testing for the product's successful prediction.

Now let us consider the content of this paper.

> The vibration facility will be capable of testing satellites of over seven tons using two Data Physics SignalForce LE-5022-3 water-cooled electrodynamic shakers.
>
> To realistically simulate the mechanical loads experienced during a rocket launch, one shaker will operate in the vertical plane (Z axis), while the other will operate horizontally (X and Y axis). Each shaker is capable of developing up to 50,000 lbf (222kN) of force and has up to 3 in (76.2 mm) peak-peak displacement. One of the key features of the LE-5022-3 is the use of dual hydrostatic bearings for axial guidance and cross-axial restraint.
>
> The horizontal system will employ a slip table incorporating Team Corporation's unique T-Film bearing system, which offers the rugged overturning moment capacity of the Team hydrostatic bearings combined with the full low-pressure oil support of a granite of journal bearing oil film type slip table. This table system is assembled from a number of patented modular T-Film bearings. Each bearing measures 12" X 12" and allows the table to be configured to the customer's test article size and load requirements. The use of multiple bearings to completely support and guide the slip plate will offer excellent dynamic stability and damping for better test results....

The above statement has one positive methodological aspect—it replaces the system that was used by the automotive industry many years ago. Of course, the design details are new. But it has the following methodological disadvantages, for example, in comparison with automotive industry:

- The vibration acts only in two degrees of freedom—vertical and horizontal. However, as has long been known in the automotive and other mobile industries, it must vibrate in six degrees of freedom—three linear and three angular.
- The author visited the Jet Propulsion Laboratory (JPL), Langley Research Center, and other aerospace research centers working on testing issues. He asked to be shown their reliability testing equipment and components of this testing. It could not be done. In that time the automotive and other mobile equipment had this fault. A similar situation can be seen now in the aerospace industry.
- Therefore, the above "developing of testing capabilities" clearly demonstrated how the aerospace area testing technology (methods and equipment) is behind the other areas of mobility.

No significant difference from the above problems are seen in the contents of other issues *Automotive Testing Technology International*. In the September 2021 issue, Steffen Schmidt wrote in his article "Simulation vs Real World" [28A]:

> Imaging the future of vehicle development without simulation is no longer an option. Simulation is a key facilitator to mastering the rising test efforts caused by more complex and connected vehicles. Whenever objective measurement results are required, virtual test-

ing comes into play. Plus, simulation can be used seamlessly in the entire development process, even at the early development stages when no real prototypes are available.

Particularly with regard to the growing importance of ADAS, autonomous vehicles and their validation simulation is playing a critical role. Countless scenarios need to be tested, including traffic simulations that can be a source of danger to people and material. Due to the advanced performance of sensor simulation, its use is already common today.

Simulation also contributes significantly to the latest transformations on the development processes, for example, to enable over-the-air software updates. The integration of the system under test into a virtual prototype enables continuous, agile software development and validation between the quality gites. Virtual prototypes are used to test driving functions over millions of miles highly parallelized on HPC systems or in the cloud.

The growing importance of virtual testing due to current developments in the automotive industry is obvious. Results from real tests, however, are still necessary to validate simulation results.

The challenge in simulation is the objectification of human perception—it is not always possible to calculate how the human would assess the interaction with a car and its control systems.

Here, the vehicle-in-the-loop evaluation method offers an ideal opportunity to combine the advantages of reality and the virtual world: the driver can physically experience the test and the vehicle reaction by driving a real prototype in a simulated environment.

As we can see from the above:

- The article considers only predesign stages of development.
- The article relates to virtual world simulation and not to real-world simulation.
- The article considers only autonomous vehicles.
- Finally, as the above articles, nothing new in field conditions simulation has been considered. Also, the content does not correspond with the title.

Let us consider one more source—Allison Transmission's new Vehicle Environmental Test center in Indianapolis [29] where:

The 60,000-ft^2 VET, as it's known, is part of Allison's $400 million investment in engineering resources and new products this decade, including a new 95,000-ft^2 Innovation Center also based on the Indy headquarters campus. The VET is unique among US Midwestern vehicle test facilities (and rare in North America) in having what some customers have called "an ideal combination": two chassis-dynamometer-equipped environmental chambers capable of generating temperature extremes of between -54 degrees and 125 degrees F; full simulation of altitudes up to 18,000 ft (5,486 m) and duty cycles that are vital for regulatory compliance, on-board diagnostic (OBD) development, and vehicle performance optimization.

The CFR 1066-compliant (Code of Federal Regulations, the required test procedures for measuring exhaust, evaporative, and refueling emissions) test cells can conduct performance, emissions, and fuel consumption testing of alt fuel, hybrid, and electric vehicles and is planning to add hydrogen fueling and fuel cell vehicle (FCV) capabilities.

Further Brad Stamper, program manager at Allison Transmission, noted:

"With the VET, development teams don't need to take a two-week road trip to get three days of testing". "We can get any environment we want here. Our goal is 'road replication': Instead of taking five different trucks in five configurations on a test trip somewhere, we can use one truck, simulate the different configurations, and get it all done in a week. We bring a laboratory environment into the various development stages" [29].

Test World has lifted the covers on a new ASTM Traction Truck, an all-electric test vehicle designed to conduct ASTM 1805 tire testing for customers from all over the world [30]:

> The Test World ASTM Traction truck is designed for testing tires according to the ASTM 1805 test standard for single wheel driving traction in a straight line on snow-covered surfaces. The standard helps drivers to identify tires that provide a higher level of snow traction. Tires meeting the standard are branded with the three-peak mountain snowflake (3PMSF) symbol.
>
> The test method employs an instrumented, rear-wheel drive test vehicle—the Test World Traction Truck—to measure average longitudinal and vertical forces acting on the test tire under acceleration.
>
> Not only is Test World the first European provider to offer ASTM 1805 testing, but its Traction Truck is also the world's first all-electric iteration of such a test vehicle. The electric power train setup is ideally suited to manufacturers looking to test their tires in a more environmentally friendly way, and in-line with shifting market trends. The electric drive truck also provides more test options over the traditional internal combustion engine vehicles, as it is easier to fine-tune the test parameters for different type sizes.
>
> Test World's parent company, Millbrook, is merging with French company UTAC CERAM. The merger benefits Test World's customers as the group now boasts an even more comprehensive "tire testing" capability. Test World's year-round snow and ice testing will be complemented by the UTAC CERAM Millbrook laboratory tests and soon-to-be opened proving ground in Morocco [30].

As we can see from the above, early mentioned negative trends is transferring now to electrical vehicles.

Figure 5.11 demonstrates the common trends of publications in the development of accelerated testing in the automotive industry: the technological testing development without drivers is climbing, while the reliability and durability testing of AV is going down. This trend is similar to compared in Fig. 5.10 negative trends in virtual and physical testing.

Fig. 5.11 The common trends of publications in development of accelerated testing for the last 30–40 years (1, reliability and durability testing; 2, autonomous vehicle (AV) technology testing (without driver)

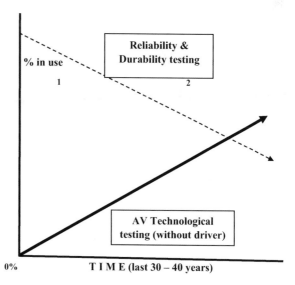

As was written in [1–4], the author's approach to development successful efficiency prediction and its basic components—methodology and ART/ADT—relates not only to mobile equipment but also to other areas of people's activity.

The content of Figs. 5.4, 5.10, and 5.11 are interconnected. Considering all of them, the reader will better understand the meaning of accelerated testing development negative aspects.

Let us review one more area—nuclear weapons. Below we can read the following improvement of testing in this area [31]. In nuclear weapons development, as in others complicated systems, traditionally engineers have separately simulated and tested each environment that weapons systems and components would experience—vibration, shock, spin, and inertial load. Over the years, Sandia National Laboratories have advanced these tests, developing capabilities to test more than one environment simultaneously.

A team of Sandia National Laboratories engineers developed a new testing capability in support of its nuclear weapons mission. The team completed their 'first combined environments' test on a full-scale weapons system at the Sandia Superfuge/Centrifuge complex in Albuquerque, New Mexico.

In a successful test, weapons engineers simulated three environments—acceleration, vibration, and spin—simultaneously on an inert experimental test system built by Sandia and utilized in collaboration with Lawrence Livermore National Laboratory. In a laboratory setting, the test created the harsh environments that weapons systems experience from launch to reentry through the atmosphere.

Sandia is the design and engineering lab for most nonnuclear components in the US nuclear weapons stockpile. The Nuclear Security Enterprise relies on Sandia for its sophisticated tests and computer models to qualify nonnuclear systems under its stockpile stewardship role. The labs' work to modify and upgrade the stockpile through life extension and modernization programs ensures its safety, security, and reliability.

And

"For the past decade, we've been running superfuge tests at Sandia, combining multiple environments. But we've really only done these tests on individual components and subassemblies," explained Paulina Rabczak, an engineer from Sandia's California laboratory working on the project. "We've now successfully designed and built an extensive, large test fixture to support testing a full weapons system and put it through flight-like environments at the superfuge," she said. "This is possibly the closest we can get to replicating an actual flight reentry event on the ground" [31].

By reproducing a flight environment in a lab setting, engineers can achieve test repeatability, further improving the reliability of test data and maturing hardware, according to Rabczak. This is expected to lead to decreased qualification time and associated costs.

The newly developed test, completed under NASA's Office of Engineering and Technology Maturation, produces richer data and better insights and is a crucial step in advancing the qualification testing of weapons systems, explained by Rabczak. Qualification tests are used to validate weapons design and systems performance.

Collecting data from a superfuge test is technically challenging due to the large amount of it and the presence of multiple spinning joints. Engineers use the data collected from environmental testing to inform and improve their designs. To support the data collection requirements for the full-system combined-environments test, engineers developed an onboard data acquisition system, which collected data from more than 200 sensors on the test unit.

The data collection system is designed jointly by telemetry and systems engineers at Sandia, rode along with the test unit, digitized the analog data close to the test, and encoded the data using established protocols. Engineers were then able to pass hundreds of channels of digitized data down to the centrifuge arm across a slip ring, a device that allows the transmission of electrical signals between stationary and rotating structures. The method would not have been possible using analog data [31].

Comparing the above statement to the fast development of electrical vehicles (EV), there are trends in EV/batteries that merit attention. Of the many companies testing equipment, the more reliable one is Weiss Technik, which developed an environment test chamber solution for the EV/battery [32]. Its battery testing equipment and battery testing systems play a pivotal role in the development and production of computer equipment, smartphones, and electric vehicles. By utilizing Weiss Technik's battery testing equipment and battery testing systems, one can calculate its approximate productive life. But one cannot successfully predict the durability of test subject during service life, because this equipment has simulated only several from many real-world environmental components.

But this is not Weiss Technik's mistake. This is a result of poor customer standard requirements.

Why are they doing so? The basic reason is saving money for testing processes. This author here repeats his previous publication [3] used in the citation of Phillip Coule; the former director of the DoD Operational Test and Evaluation Office testified in the American Senate that during the design and manufacturing of the complicated apparatus, one tries to save a few pennies in testing, and the end results may be a huge loss of thousands of dollars due to faulty products that have to be replaced because of this mistake. This problem is reviewed in this book.

Weiss Technik manufactures battery testing equipment and battery testing systems that simulate environmental influences under accelerated conditions in corresponding with current standards. The Weiss Technik's test chambers simulate the following conditions:

- **Temperature:** The temperature shock test chamber assesses the effects of extremely rapid temperature changes ranging from −80 °C to +220 °C.
- **Humidity:** The temperature-humidity test chamber can reach 98% humidity (RH) at 95 °C.
- **Tensile strength:** Test equipment can ensure the quality of materials and integrity of design and construction under realistic conditions.
- **Dust:** Equipment inject compressed air through special nozzles and add dust to create a swirl.
- **Water** splash: To determine if housings are protected from water entering.
- **Vibration:** Equipment measures the stress on parts caused by intense vibrations during use.
- **Altitude:** Test chambers can simulate heights of up to 100,000 ft for use of batteries used in aerospace and aviation.
- **Vacuum:** Thermal vacuum chambers accurately reproduce outer space conditions for pressure, cold, solar radiation, ultraviolet (UV) light, and cosmic rays.

- **Stress:** Highly accelerated life test (HALT) and highly accelerated stress screening (HASS) tests exist in these chambers.

As we can see from the above, the simulation conditions are advanced in comparison with many other companies and correspond to current standards. However, these standards do not require reliability or durability testing and consequently cannot help to obtain accurate initial information for successful efficiency prediction.

As was mentioned previously in this book, even using HALT and HASS does not help accurately simulate real-world conditions.

We reviewed before the dedicated publications on accelerated testing methods and technologies.

Let us consider next the program of Electric & Hybrid Aerospace Technology Symposium, May 31–June 1, 2022, Frankfurt, Germany.

The topics under discussion included the following:

- The possibilities created by aircraft hybridization
- Commercial aircraft application possibilities and research
- eVTOL propulsion and aircraft design
- Battery technologies
- Electric motor technologies
- High-speed engine and turbofan technologies
- Road world fuel saving possibilities
- Energy storage systems
- Increasing flight range through hybridization
- The possibilities of pure-electric only
- Safety and legislative considerations
- Case studies on existing global electric and hybrid research programs
- Overcoming engineering challenges
- Best design practices
- Investment possibilities
- Additional advantages of increased electrification
- Range extender technologies

As we can see from the above list of topics, the conference dedicated no time for testing and efficiency prediction development. Rather, it demonstrates where the attention of designers, researchers, and manufacturers is being given. Their ideas about accelerated testing development are as follows:

- Most current publications in testing development, especially in 2019–2022, concentrate on autonomous vehicles (AV) consideration.
- These publications relate mostly to technological testing—considering how the vehicles are working without drivers.
- Many other publications relate to vibration. They demonstrated new details of vibration testing, but nothing new in development vibration technology (methods and design).
- Reliability and durability testing development is not underway (the reasons were considered in this book).

- The level of sophisticated and more effective testing for obtaining initial information for product's successful prediction is not improving.
- Current simulation procedures reflecting real-world conditions is not accurate, which leads to poor level of testing and efficiency prediction.
- It was similar to a situation when 5–10 years ago physical testing jumped to digital testing. The author analyzed that situation in his book *Trends in Development of Accelerated Testing for Automotive and Aerospace Engineering,* Elsevier, 2020 [23].
- In the nuclear area of development, there is current positive trend—Sandia National Laboratories began to use simulation and testing simultaneously on four "environments—vibration, shock, spin, and inertial load." But this level of simulation and testing is lower than reliability or durability testing, because these several types of environments are not accurate and reflect the real-world situation, where there are many types of environments.

As was mentioned, less attention to testing, quality, reliability, and other efficiency components is a negative trend in the development of new prediction technology where it is such a critical part of transportation, aerospace, and defense equipment. For example, let us consider the SAE International magazine *Aerospace & Defense Technology,* where we see that their articles during 2022 concentrate on electronics related to new product development with nothing on reliability improvement. To illustrate, here are the contents of the May 2022 issue [37]:

- Powering Better Battlefield Drones Using Low-Frequency Broadband Sonar on UUVs
- Designing Rugged SWaP-Optimized MOSA Solutions for UUVs
- Does Your UAV Program Need a Transponder?
- Developing New Anti-Drone Radar Technology
- Deceiving the Enemy: These Are the Drones You Are Looking For
- Calibrated Long-Wave Infrared (LWIR) Thermal and Polarimetric Imagery of Small Unmanned Aerial Vehicles (UAVs) and Birds
- Cyber Risk Assessment and Scoring Model for Small Unmanned Aerial Vehicles
- Modeling a UAV-Based Mesh Network to Analyze Latency and Throughput
- Investigation of Flight Dynamics and Controls for a Solar-Tracker-Mounted UAV

One more example of negative trends in development reliability and durability testing. In the [38] was written the common description of these types of testing without their specific requirements. In the EV testing measurements, "Follow Your Data" was written [38].

Follow Your Data: How to Effectively Get Insights from EV Testing Measurements

Electric Machines: Increase Reliability
Learn how to enable engineers to understand and confirm how a machine holds up to real-world use by entailing long-term electrical machine testing, typically carried out for a defined minimum period. Read more about durability testing, its benefits, and HBK solutions.

Evaluate and Monitor Motor Durability

Electric motors and power train present new challenges for durability testing, such as new failure modes in motors like demagnetization, delamination, and turn-to-turn shorts.

Check out this video which presents:

- Failure modes of electric motors
- Benefits of measuring electrical values for durability testing
- Real data from durability testing

How Durability Testing Increases Reliability of Electric Machines

Durability testing, an essential stage in the development of any electrical machine, enables engineers to understand and confirm how a machine holds up to real-world use. This entails long-term electrical machine testing. Typically, durability testing is carried out for a defined minimum period to ensure the product survives without failure or for however long it takes for the machine to break down and repairs are necessary.

How to Determine Durability

Durability measures the length of a product's life. There are several measures of durability in use, including years of life, hours of use, and number of operational cycles. Reliability testing ensures that product quality is consistent with its specifications throughout the intended life cycle. This testing can be done at both the design and production levels: to identify and mitigate flaws in design or to correct potential failures in manufacture.

Reliability Testing with HBK

The HBK Durability Testing solution validates and verifies the properties and functions of electrical components and sub-systems. During the test procedure, "condensed" parameters are recorded, like RMS, P, η, λ, and ρ. Measurements are taken according to a time grid for the test at intervals that may range from once per several seconds to multiple times per second. Raw data can be recorded either in fixed time intervals for a short period (to understand the "aging" of the motor") or based on trigger conditions (e.g., a motor breakdown, and to understand this with pre-trigger), or both.

In addition to recording measurements, Perception software—used to analyze the results of durability testing—has a database of standard formulae that can be calculated and stored in real time. The software is easily adapted to custom configurations and can be extended with additional user-specific formulae as required.

Automotive Durability Testing

In the automotive industry, durability testing supports the component and OEM manufacturers' evaluation of vehicles to determine the expected service life of components or sub-systems. These assessments are made at the test bench by replicating real-world conditions and stresses that the vehicle would experience through actual use. Variations on the terminology include accelerated vehicle testing, full-vehicle durability testing, high-performance durability testing, and so on—but each follows the same basic principles.

Basic Requirements for Durability Testing

Typical requirements for measurement configurations are the following:

- A sample rate of 100 to 200 kS/second
- Measurement and data collection from 6 to 18 power channels simultaneously, with voltage up to 1000 volts and current up to 1000 amps
- Single-drive setup, motor-generator setup, or more complex like eCVT
- Additional signals such as temperature and/or vibration
- CAN bus inputs
- Transfer of results to the test automation system

As we can see from the above, nothing shows about how one can do it. Nothing about basic requirements to these types of testing—accurate simulation of real-world conditions. Therefore, the given requirements are poor. How do the authors understand the term "real-world conditions"? As was mentioned in this book, if one calls testing as "reliability testing" or "durability testing" without basic requirements to these types of testing, this negative trend leads to obtaining results that are different from real-world results. And these testing results lead to inaccurate prediction of reliability and durability of the test subject, as well as other components of the product efficiency.

One more example. If we consider the program and exhibition of Aerospace Testing Symposium [39], we will see mostly flight testing and laboratory testing of details. There's nothing about reliability or durability testing.

As we can see from [40], Volkswagen near Chattanooga, Tennessee becomes the first non-domestic OEM to fully integrate battery test and validation, pack assembly, and EV manufacturing in the US [40]. In this work were involved so famous companies as MTS, Weiss Technik North America, and others.

We are transforming this region into a powerhouse for EV production, including suppliers, asserted Johan de Nysschen, Volkswagen, Group of America's COO, during a tour of the BEL in June 2022 [40].

If we will analyze how they plan to provide simulation and testing technology, we will see considered above negative trends of separate: thermal shock chamber, drive-in temperature chamber, dust chamber, water immersion and sodium chloride (salt-bath tests that "simulate up to one year of EV life in a hostile winter environment, such as southeastern Michigan, and a water-immersion test that simulates water ingress after shocking the aluminum case structure, are keystones in VW pack development.

This contradicts the real-world, where these procedures act simultaneously and in combination. Why are they doing so? The basic reason is saving money for testing processes and not thinking about loss of money during subsequent processes.

References

1. Lev M. Klyatis (2012) Accelerated Reliability and Durability Testing Technology. Wiley.
2. Lev M. Klyatis, Edward L. Anderson (2018) Reliability Prediction and Testing Textbook. Wiley.

3. Lev M. Klyatis, Eugene L. Klyatis (2006) Accelerated Quality and Reliability Solutions. Elsevier.
4. Lev Klyatis (2016) Successful Prediction of Product Performance. quality, reliability, durability, safety, maintainability, life-cycle cost, profit, and other components. SAE International.
5. Lindsay Brooke (2021) Making the Strategic Move Into AV Testing. Automotive Vehicle Engineering. July 2021.
6. John Wilson (2018) Why So Many Product Recalls? TEST Engineering and Management. April/May 2018.
7. Stuart Birch (2010) Mercedes' CLS: is most tested car. Automotive Engineering. SAE International, October 2010.
8. Mike Pendleton (2021) Beyond Battery Cycling: Testing Batteries in the Automotive Industry. SAE Webinar.
9. Sarath Jayatilleka, Geoffrey Okogbaa (2014) Accelerated Life Testing (ALT). ASTR. 2014 Workshop on Accelerated Stress Testing and Reliability.
10. Kyle Hailey, ITworld | May 15, 2015. Poor Test Data Costs Industry Billions Per Year. *https://www.itworld.com/article/.../poor-test-data-costs-industry-billions-per-year.html*
11. John P. King and William S. Jewett (2010) Robustness Development and Reliability Growth: Time, Money, and Risks. Premtice Hall. INFORMIT. April 2010.
12. William Q. Meeker and Luis A. Escobar (1993) A Review of Recent Research and Current Issues in Accelerated Testing. *International Statistical Review / Revue Internationale de Statistique.* Vol. 61, No. 1, Special Issue on Statistics in Industry (April 1993).
13. Symposium "Autonomous Vehicle Test & Development" (2018) (North America's only conference dedicated to test & validation of autonomous vehicles & self-driving technology). October 23–25, 2018 in Novi, MI, USA.
14. Autonomous Vehicle Test and Development Symposium in conjunction with Autonomous Vehicle International magazine. 5 - 7 June 2018, Messe Stuttgart, Germany.
15. Automotive Components Fatigue and Durability Testing with Flexible Vibration Testing Table **10-02-01-0004.** SAE International Journal of Vehicle Dynamics, Stability, and NVH-V127-10EJ.
16. Praveen Kumar, Prakaash J. Kumar P. (2018) Optimization of Proving Ground Durability Test Sequence Based on Relative Damage Spectrum. SAE Paper 2018-01-0101.
17. Presead S., Prakaash J. and Dayalan P. (2017) Study the comparison of Road Profile for Representative Patch Extraction and Duty Cycle Generation in Durability Analysis. SAE Technical Paper 2017-26-0309, doi:https://doi.org/10.4271/2017-26-0309.
18. Radar Reliability Testing (2019) The 14th Institute of the China Electronics Technology Group Corporation. Aerospace Testing International. June 2019. Page 83.
19. Automotive Testing Expo 2021, Novi, MI. October 26, 27, 28, 2021, Booth 6002.
20. James Truchand (2008) The Emerging Role of Physical Test in Product Development. SAE International. Automotive Engineering.
21. Back to Environmental Basic (2014) Aerospace Testing International. April 2014.
22. Lalanne C. (2002) Fatigue Damage. Mechanical Vibration and Shock. Volume 4, ISBN 1903398066, April 2002. [22A]. MTS Systems Corporation. www.mts.com
23. Lev M. Klyatis (2020) Trends in the Development Accelerated Testing for Automotive and Aerospace Engineering. Academic Press (ELSEVIER).
24. Automotive Testing Technology International, November 2020.
25. Danny Atsmon (2020) CEO, Cognata. Data Farming. Automotive Testing Technology International. November 2020., pp. 048–052.
26. Building Modular Test Centers (2020) Compared With a Traditional Block-Built Test Center building, modular test facility solutions save a lot of time and trouble. Automotive Testing Technology International. November 2020. pp. 091–093.
27. SAE Aerospace International. November 20, 2020. [27A]. Dave Baker (2020) How to Future Proof Your Test Systems. Aerospace Testing International. SHOWCASE.

28. Kevin W. McIntosh and Thomas Reilly (2020) Developing a National Satellite Testing Capability. Aerospace Testing International. ShowCase. November 2020. [28A]. Steffen Schmidt 2021 Simulation vs. Real World. Automotive Testing Technology International, September 2021.

29. Lindsay Brooke (2021) Allison Builds a Testing Powerhouse. Truck and Off-Highway Engineering Technology Newsletter. 2021-02-09SAE International. 2021-02-09

30. News 2021. Test World Unveils World's First All-Electric ASTM Traction Truck Test World Unveils World's First All-Electric ASTM Traction Truck MARCH 10TH 2021 PROVING GROUNDS,WINTER DRIVING, FINLAND

31. New Weapons Testing System Produces Richer Data, Saves Time and Money (2021) Defense E-Newsletter, September 2021. newsletters@newsletters.techbriefsmedia.com

32. Weiss Technik. Safe Automotive Battery Testing with Environmental Test Chambers. Environmental Test Chambers Solutions EV/Battery.

33. DoD 3235, 1 H. *Test and Evaluation of Systems RAM* .

34. IEEE 1413-2002: *A standard Methodology for Reliability Predictions. Guide for Selecting and Using Reliability Predictions based on IEEE 1413. 2019* [7].

35. MIL-HDBK-217F. *Military Handbook. Reliability Prediction of Electronic Equipment.*

36. Military Standard MIL-STD-781D *Reliability Testing for Engineering Development, Qualification, and Production.*

37. SAE International magazine *AEROSPACE & DEFENSE TECHNOLOGY.* May 2022.

38. Hottinger Bruel & Kjaer news@news.hbkworld.com , 5/19/2022, 2:12 PM. Electric Machines: Increase Reliability

39. Aerospace Testing Symposium. Mark Allen Group. September 27–28. London. UK.

40. Lindsay Brooke. Volkswagen sprints forward in the EV race by integrating battery test, production, and EV assembly in the US. Automotive Engineering. SAE. September 2022

Chapter 6
Implementation Successful Prediction of Product Efficiency, Accelerated: Reliability and Durability Testing, and Accurate Simulation

6.1 Introduction

As was published in [1–6] and others, the author developed useful means for practical implementation of a new scientific/technical direction of efficiency successful prediction and its basic components as follows:

- Methodology of this prediction
- Technology for accurate simulation of real-world conditions
- Source for obtaining initial information for successful efficiency prediction—accelerated reliability and durability testing technology

One can see in this chapter how the implementation of some of these directions comes from components of different disciplines (engineering, physics, and mathematics) being used in different countries.

Some of these implementations were connected with the author's career and increased after his immigration to the USA. Below (Fig. 6.1) one can see a calendar of these activities. Of course, a successful immigrant's career cannot be without investments of ideas, time, and effort.

Here are several forms of such investments:

- Development of innovative ideas, technology or methodology, and their implementation for positive results
- Worldwide leadership in new scientific/technical strategy development
- Solutions of important problems in reliability, efficiency, and productivity
- Money invested for development of research projects

Before coming to the USA, the author had accomplished the first three forms of investment, but communicating them was a difficult problem. Of course, his ideas

---

Supplementary Information: The online version contains supplementary material available at https://doi.org/10.1007/978-3-031-16655-6_6.

193
L. M. Klyatis, *Prediction Technologies for Improving Engineering Product Efficiency*, https://doi.org/10.1007/978-3-031-16655-6_6

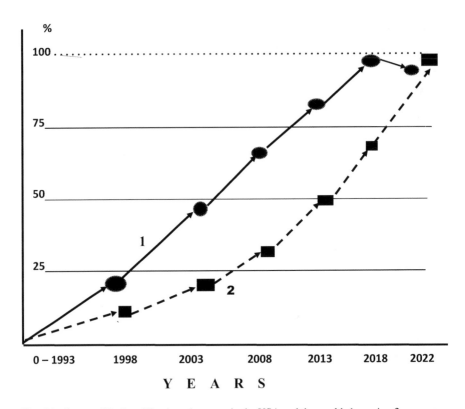

Fig. 6.1 Career of Dr. Lev Klyatis and renown in the USA and the world: 1, carrier; 2, renown

were familiar to a number of Russian professionals. However, English is the international language and the author was not familiar with it when he came to the USA. Therefore, he had to overcome three basic obstacles to obtain the possibility to implement his ideas and technology:

1. Learn the English language, which was difficult as the author was not so young.
2. Through English he can introduce his ideas and solutions through presentations and publications and directly contact professionals of many nationalities.
3. This necessitated money for traveling to symposiums and world conferences. At first, earning money for living as a fish delivery took priority as seen in Fig. 6.19. Eventually, he got a research grant award from the American Society for Quality (ASQ). The author was financially assured that with his work in West Germany and in the USA with Soviet Union governmental delegations along with visiting world leader companies located in Russia, Belorussia, and Ukraine, he would be able to provide scientific/technical solutions to problems not yet solved in Western, Asian, and other countries. This completed the way for the author's investment in successfully developing and implementing his ideas and solutions

Career "steps" after coming to the USA:

1995–1997	Updating ASAE (American Society Agricultural Engineers), standard ASAE Engineering Practice: ASAE EP456 Test and Reliability Guidelines
1996–2002	Seven published presentations at the Annual Reliability and Maintainability Symposium (RAMS)
1998	Obtained Research Grant Award ($ 10,000) from American Society for Quality (ASQ) (8 annual awards for 120,000 members)
1999–2001	Reliability division manager for ASQ Annual Quality Congresses
1999–2001	Consultant Reliability Analysis Center; Department of Defense
2000–2005	Expert reviewer in reliability for publisher ASQ Press
2000–2005	Accredited delegate of the US Technical Advisory Group and representative of US National Committee for International Electrotechnical Commission (IEC) for international standardization, Technical Committee TC 56 "Dependability"
2000–2005	World Quality Council member
2000–2019	SAE International Technical Committee G-11 Reliability, Maintainability, Supportability, and Logistics member in aerospace standardization
2003–2004	Consultant for the US Agency USDI/VOCA for helping undeveloped countries
2004–2005	Expert subsequent nominated to IEC/ISO Joint Study Group ACOS (Safety Aspects of Risk Assessment), Frankfurt am Main, Germany, for international standardization
2005–2013	Developed six SAE International standards in aerospace engineering (reliability testing, JA1009)
2002–2020	Published seven books in successful prediction of product performance, accelerated reliability and durability testing, accurate physical simulation of real world conditions by world publishers Wiley, Elsevier, Academic Press, SAE International, and Chinese Defense Publishing
2004–1015	Consultant for governing organizations and international companies, including Ford Motors, DaimlerChrysler, Toyota, Nissan, Jatco, Black and Decker, John Deere, in efficiency prediction, accelerated reliability/durability testing, and accurate physical simulation of real-world conditions
2006	Elected American Society for Quality (ASQ) fellow
2007	Elected SAE International fellow
2007	SAE International representative for Elmer A. Sperry Board of Award
1995	Implemented research results of sciences and practices in many engineering areas and countries
2012	In the SAE World Congress in Detroit was included special technical session "Trends in Development Accelerated Reliability and Durability Testing" (continued in other years' congresses), where Lev Klyatis was invited as a chairman
2015	Advisory Board Member (design, manufacturing, and testing); Cambridge Scholars Publishing
2007–2020	Seminar instructor and lecturer for societies, governmental organizations, universities, and companies
2021	Adviser SAE International association
2021	Author's book *Trends in Development Accelerated Testing for Automotive and Aerospace Engineering*, published by Academic Press (2020), as featured on Forbes and CNN and rated among the 100 best aerospace and automotive engineering books for all times
2010–2020	Lecturer at American universities
2022	Continuation work with presentation at the SAE 2023 World Congress, SAE G41 Standardization Committee, preparation of new book for Springer Publisher, co-organizer for IDM100 SAE 2023 World Congress, published paper for SAE 2022 World Congress, Elmer Sperry Board of Awards

6.2 Common Principles

While there are diverse types of implementation, all of them must begin through the brains of the students and professionals involved in the study of the innovative approaches, technologies, strategies, methods, and the equipment needed for their practical implementation.

Therefore, without, first, studying, learning, and teaching all those involved in the research, design, testing, manufacturing, service, and usage of the product are minimally effective. For accurate real-world simulation, reliability and durability testing, and successful efficiency prediction implementation, one needs to understand the components of this prediction and how it can be done on a step-by-step process. Included in this process, one must accurately gather sample measurement data that reflects the real world that is confronting the equipment being tested.

The fundamentals necessary for implementation of latest ideas, strategies, approaches, and technologies can be accomplished through the following:

1. Analyze publications detailing the latest ideas, approaches, and technologies in the literature, including various books, articles, papers, lectures, published presentations, dissertations, and reports. After this, it is very important to accept the new concepts advanced in this literature, to be used by the reader's practice for improving results of his or her job.

2. Learning about the innovative ideas, approaches, and technologies from learned lectures and presentations. Especially important for effective implementation is hearing or reading expert author's innovative ideas in their lectures, tutorials, presentations, seminars, or consultants.

3. Using this author's citation from publications of innovative ideas, approaches, or technologies.

4. Using citations of the author's basic ideas and research results from his books, articles, papers, and project plans that provide an important level of latest ideas for implementing original approaches and technologies.

5. Special lectures for future engineers, managers, and top management of companies directing implementation of new strategies.

6. Including the knowledge from these approaches and technologies in the preparation of appropriate international, national, and industry standards.

7. Implementation of innovative ideas after discussions with expert professionals.

8. Upper-level management buy-in obtained to provide presentations, lectures, and tutorials detailing the original approaches, technologies, and the relevant citations with their author's names, including their benefits to their organizations and customers/users.

9. Develop new organizational structures, whether through educating existing groups or by forming new teams, divisions, departments, or companies who will be responsible for developing and implementing the latest ideas and technologies.

10. Implementing original approaches and technologies and documenting the results thereby demonstrating the improvements gained and economic results through this implementation of new ideas and technology.
11. Document and disseminate the economic gains to the organization achieved by using the new ideas, approaches, and technologies.
12. Mentioning the author's book in Forbes, CNN, and other famous media, as the best books for establishing new implementations of his strategies.

The author created his ideas of, first, successful reliability and durability prediction while working in the USSR, where his methodologies and tools for this new scientific direction were developed.

Then he continued his work in the field after moving to the USA, continually improving his strategies and methodologies.

As will be shown below, this implementation process is continuing today in different countries around the world.

For example, in 2020–2022, the implementation process was continued through the following:

- Lecture "Introduction to Accelerated Reliability/Durability Testing: Reliability and Safety Accurate Prediction and Successful Problems Prevention" at the New Jersey Institute of Technology, February 2020 (Fig. 6.2).
- SAE 2020 World Congress paper (#2020-01-0533) and presentation of "Analysis of Basic Directions of Accelerated Testing Development." One can find now this paper in many universities and public libraries (Pennsylvania University Libraries; Cleveland Public Main Library; University of Liverpool, Sydney Jones Library, UK; University of Canterbury, Central Library, New Zealand; and others).

- SAE 2021 World Digital Summit published the author's paper 2021-01-0236 "Implementation of Positive Trends in Reliability and Durability Testing and Prediction," where the specifics of positive trends were reviewed and how they are implemented in practice.
- SAE 2022 World Congress published the author's paper 2022-01-0212 "Analysis of the Current Trends in Development Real-World Accelerated Testing," where he showed how the vehicle testing development technology is failing behind more than going forward and how one can improve this situation through increased use of accelerated reliability/durability testing (ART/ADT) technology.

The author developed new direction of successful efficiency prediction, and its components are used in many countries over the world. For example, the author's book [1] was translated in Chinese [7] and published in China. One can see the front cover of this book in Fig. 6.3.

Fig. 6.2 Dr. Lev Klyatis is providing the lecture "Introduction to Accelerated Reliability/ Durability Testing: Reliability and Safety Accurate Prediction, and Successful Problems Prevention," New Jersey Institute of Technology, February 2020

6.3   Direct Implementation: Economic Results

The first successful direct implementation of product's efficiency theories was made in Russia (Kalinin State Test Center and company Bezeckselmash) for farm machinery. Later, some of these ideas, methodologies, and new test equipment for this were adopted in the Ukraine, Zelenograd City, Moscow State (see equipment for vibration testing in six degrees of freedom in Fig. 5.40 [1]), Israel (Figs. 6.3 and 6.4), and Belorussia (Fig. 6.6), and then, from there, components of these prediction approaches and methodologies were implemented in other countries.

This book's approaches to accelerated reliability and durability testing (ART/ ADT) and product successful efficiency prediction have been implemented by many industrial companies. As can be seen by the example (company in Israel) in Fig. 6.4 and Tables 6.1 and 6.2, there are many economic benefits to these approaches.

Until today, one major obstacle for implementing these innovative approaches is how many professionals in the areas of design, testing, production, and marketing of complex product, both for the companies and their suppliers, worked in isolation without sufficient interaction among themselves. They worked with separate requirements (standards) and responsibilities that often are not connected with each other. This is the first problem companies need to overcome for successful implementation of these approaches.

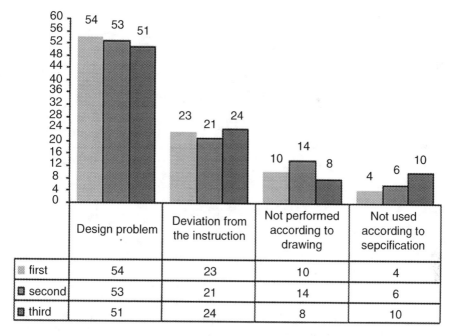

	Design problem	Deviation from the instruction	Not performed according to drawing	Not used according to sepcification
first	54	23	10	4
second	53	21	14	6
third	51	24	8	10

Fig. 6.3 Complaints and their reasons during 3 years after implementation of the new approaches for successful prediction for the tools (Israel)

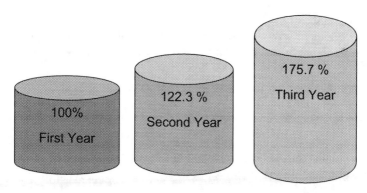

Fig. 6.4 Increased volume of sales of instruments after the implementation of the innovative approaches for successful prediction (Israel)

The new approach shows that how the team members need adequate knowledge in metallic details, new composite materials and units, etc. and high-level standardization, especially in quality, reliability, maintainability, durability, and prediction.

The following are practical examples that illustrate the superior possibilities of the author's approach to ART/ADT for the rapid improvement and prediction of product quality, reliability, durability, and cost-effectiveness over their products life cycles.

Table 6.1 Comparison of the complaints for a reason

| Reasons | % of complaints | | |
| | Years of implementation | | |
	First	Second	Third
Design problems	58	43	36
Deviation from the instruction procedure	23	31	34
Not performed according to drawing	10	10	8
Not used according to specification	4	10	15
Others	5	6	7
Total (%)	100	100	100

Table 6.2 Example of practical economic results of the proposed approach for tools

Year	Sales (Million $)	Complaints (number)	Rejects (%)
2003	134.3	151	3.2
2004	156.2	144	3.0
2005	196.4	127	2.4
2006	214.0	114	2.0

Example 1 An industrial company had problems with the reliability and durability of a model of harvester, which they were unable to solve using existing field testing methods. They even conducted laboratory testing, but the physical simulation of the field influences in their laboratory testing was not accurate.

The chief designer of the company asked the author to use his approach for solving the problem. To accomplish this, a special methodology of ART/ADT and test equipment was developed to provide accurate prediction of the harvester's reliability during its service life.

For 6 months:

- Two of the original design harvester's specimens were subjected to reliability evaluation for the equivalent of 11 years of normal operations.
- Three variations of one unit and two variations of another unit were tested. The resulting information was equated to 8 years usage of their service life.

As a results of using these processes, the reasons for degradation process were eliminated, because:

- The harvester's design was changed according to the conclusions and recommendations drawn from the testing results.
- Specimens incorporating the design changes were then field tested.
- The manufacturing technology was improved corresponding with test results.

Example 2 There was a problem related to the reliability of the working heads some special belts that drove another model of a complex machine. The low reliability of these belts limited the reliability of the whole machine and was an issue beginning with its design stage and continuing through the manufacturing stage and in its the field use.

As an attempted remedy, the company's designers increased the strength of the belt, but this only increased reliability by 7% and was accompanied by doubling their cost. The company Bezeckselmash asked the author to help solve this problem. To do this, he created with his team test equipment that accurately simulated the field input influences on the belts. This entire process emulated the methodology described in Chap. 4. After several months of testing and studying, the cause of the poor reliability was found, and a new reliability prediction was conducted. The reasons for the belt failures were newer in the belt, but in the connector unit. From these new results, the author with the company's designers developed recommendations for improving the design of the machinery. The following field testing demonstrated that by implementing these changes, the durability in the updated machine was more than doubled. Importantly, the changes to the cost of the above work increased the machine's cost by only 1%, including the cost of the testing equipment and all the work involved in finding the reasons contributing to the failures of the belts.

One more example follows from the automotive industry. Figure 6.5 illustrates a plan for a new test chamber used for reliability/durability testing of trucks by KAMAZ, Inc. that implemented the author's strategy. Subsequently, after the author's presentation at the US-USSR Trade and Economic Council meeting, the author received a proposal from KAMAZ, Inc. (Russia) chairman Mr. N. Bekh for implementing the author's technology for developing at KAMAZ, Inc., including a new system of research, design, testing, manufacturing, and maintenance of trucks.

This was a huge offer as KAMAZ was one of the largest manufacturers in the USSR's automotive industry. Employing over 100,000 employees at that time, KAMAZ, Inc. sold its product only in the USSR and some Eastern European and other countries. But it wanted to begin selling their product in the more developed West European, American, and Asian markets. As a part of this venture into the market, they reached an agreement for Cummins (USA) to build a special plant for them in the USSR for manufacturing Cummins' engines. They planned to use the author's technology for increasing product reliability, safety, durability, maintainability, profit, and decreasing life cycle cost, thereby open the door to their new markets by increasing the competitiveness of their product in the world market with competitive product quality.

Upon returning to Moscow, the author learned from the Automotive and Farm Machinery Departments that Mr. N. Bekh was also the economic advisor for Mr. MS Gorbachev, the USSR President.

Mr. Bekh and KAMAZ's general manager Mr. Pavlov had an in-depth understanding of what successful prediction of product efficiency could mean for the success of their company. They were ready to invest many millions of dollars into this project which will bring this company billions of dollars in revenues and profit.

The result was a contract between KAMAZ, Inc. and TESTMASH, where the author was the chairman, and that included all the requirements relating to

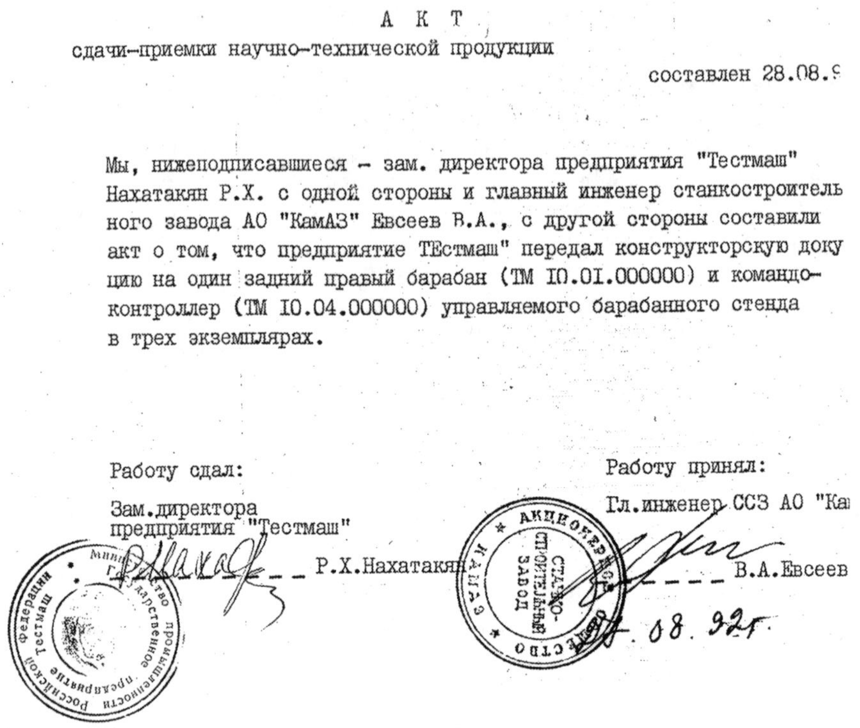

А К Т

сдачи-приемки научно-технической продукции

составлен 28.08.9

Мы, нижеподписавшиеся – зам. директора предприятия "Тестмаш"
Нахатакян Р.Х. с одной стороны и главный инженер станкостроитель
ного завода АО "КамАЗ" Евсеев В.А., с другой стороны составили
акт о том, что предприятие ТЕстмаш" передал конструкторскую доку
цию на один задний правый барабан (ТМ I0.01.000000) и командо-
контроллер (ТМ I0.04.000000) управляемого барабанного стенда
в трех экземплярах.

Работу сдал: Работу принял:

Зам.директора   Гл.инженер ССЗ АО "Ка
предприятия "Тестмаш"

_____ Р.Х.Нахатакя   _____ В.А.Евсеев

                                                . 08. 92г.

Fig. 6.5 Example of an acceptance certificate for the scientific and technical products of TESTMASH for KAMAZ, Inc.

implementation, including payment for TESTMASH's patents in several countries. With this contract TESTMASH began to design and implement solutions as required by KAMAZ, Inc. They included a specific design of a new methodology and equipment for the physical simulation of multi-environmental influences. It includes development and using a large test chamber (see plan of test chamber [TESTMASH design] for completed truck, Fig. 5.33 in [1]) that tested many interactive components, equipment simultaneously for random vibration capable of testing complete three-axle trucks, corresponding to the author's recommended methodological solutions. An example of an acceptance certificate of the TESTMASH's product for KAMAZ, Inc. can be seen in Fig. 6.5.

This was in stark contrast with author's experiences with the management of many American and Japanese companies (as described in this book) who could not understand the long-term benefits that would be achieved from successful efficiency prediction. Therefore, they were reluctant to accept his innovative ideas and approaches to reliability and durability testing and successful efficiency prediction for providing the significant economic benefits for their companies. The above companies were too afraid of a risking scarce investment money.

Fig. 6.6 Dr. Yakhya Abdulgalimov (TESTMASH) during implementation component of ART/ADT in the industrial company SELMASH (Bobruysk, Belorussia)

The test chamber design shown above incorporates the results by directly implementing the author's approach to reliability/durability and other efficiency components prediction. This test chamber includes in one complex test booth, both accurate real-world simulation and ART/ADT.

One more example. Figure 6.6 is demonstrating an ART/ADT implementation approach in SELMASH (Bobruysk, Belorussia).

With the accurate initial information from the results of ART/ADT, the methods of efficiency prediction were very productive.

The number of successful implementations such as these were increased during the author's career after his coming to the USA. One can see this situation in Fig. 6.1.

6.4 Implementation Through Standardization

6.4.1 Implementation of Reliability Testing and Successful Reliability Prediction Through the Application of Standard EP-456 "Test and Reliability Guidelines" for Farm Machinery

It began the implementation of successful efficiency prediction by employing the concept of standardization with farm machinery. We will review how this come about. The author served as a member of American Society of Agricultural Engineers (ASAE) T-14 Reliability Committee. At the ASAE International meeting in 1995, it was proposed that the author leads an effort to update the standard EP-456 "Test and Reliability Guidelines" (see Fig. 6.7). This effort resulted in updated guidelines to include research results in reliability testing and prediction (See Fig. 6.8). These guidelines were approved by ASAE T-14 Committee and then ratified by the ASAE.

One result of the author's work for American Society of Agricultural Engineers (ASAE) was an invitation to visit the People's Republic of China as a member of American delegation (Fig. 6.9).

6.4.2 How the Work in SAE G-11 Division Assisted to Implement Accelerated Reliability Testing as a Component of Successful Efficiency Prediction

The author was invited to work on the G-11 Division of SAE International Technical Committee Reliability, Maintainability, Supportability, and Probabilistic Methods. The G-11 Division prepared, discussed, updated, approved, and voted new and current standards related to the aerospace industry, which became SAE standards.

A key objective of the SAE G-11 Division was the development of standards, but these efforts were not always successful. The existing SAE draft of standard JA1009 was planned for aerospace reliability testing and prepared earlier. However, sadly, it was not approved by SAE International Aerospace Council.

The SAE G-11 Division meeting proposed that a new standard JA1009 Reliability Testing directive should be prepared. The author made presentations during the G-11 meeting and proposed a group of standards under the common title "Reliability Testing."

The resulted in a group of standards consisting of the following:

- SAE JA 1009—A. Reliability Testing: Glossary
- SAE JA 1009—1. Reliability Testing: Strategy
- SAE JA 1009—2. Reliability Testing: Procedures
- SAE JA 1009—3. Reliability testing: Equipment
- SAE JA 1009—4. Reliability Testing: Statistical criteria for a comparison of a reliability testing results and field results
- SAE JA 1009—5. Collection, calculation, statistical analysis of reliability testing data, development recommendations for improvement of test subject reliability, durability, and maintainability

This group of standards was largely based on results from the development of efficiency prediction and accelerated reliability and durability testing that were described in Chaps. 3 and 4 of this book. The SAE G-11 Reliability Committee approved the first standard SAE JA1009/A Reliability Testing: Glossary and recommended integrating this with existing standard ARP 5638.

The second standard SAE JA1009—1 Reliability Testing: Strategy was prepared, discussed, and approved for ratification, but the G-11 Reliability Committee became inactive, and final vote process was not completed.

COOPERATIVE STANDARDS PROGRAM

28 July 1997

To: Lev M. Klyatis

From: Dolores Landeck, Standards

Re: Standards Projects Status Reports

Standards Department records indicate that you are the project leader for the following standards project(s):

×456

As you know, considerable time and effort are necessary to develop, on a timely basis, standards that address industry needs, and feedback is essential to coordinate these projects. Your cooperation is therefore requested in completing the enclosed Project Status Report form and returning it to Standards staff so that we can update and verify our project records.

If possible, please fax your completed form prior to the annual meeting next week in Minneapolis, where is will be taken for review by your -03 Standards committee; alternatively, you may deliver it to me or to Russ Hahn at the meeting, if you plan on attending. If neither of these options is feasible, please return the form to the Standards Department before the end of August.

Copies of the completed form should be sent to others who need to know the status of your project and your plans for leading the project to completion—your committee chair, any project co-leaders, etc. If you have any questions, or if our records are incorrect and this memo should be addressed to someone else, please contact me by phone or email: (616) 428-6339/landeck@asae.org.

Thanks for your continued participation in the Standards program.

ASAE
2950 NILES ROAD • ST. JOSEPH, MI 49085-9659 • USA • 616/429-0300 • FAX: 616/429-3852 • E-MAIL: HQ@ASAE.ORG

Fig. 6.7 The document from ASAE Cooperative Standards Program states that the author is a project leader for the standard program ×456: the Standard EP 456 Test and Reliability Guidelines

This information was discussed and later published at the SAE 2013 World Congresses (papers # 2013-01-0152 *Development Standardization 'Glossary' and 'Strategy' for Reliability Testing as a Component of Trends in Development of ART/ ADT*" and #2013-01-0151 *Development of Accelerated Reliability/Durability Testing Standardization as a Components of Trends in Development Accelerated Reliability Testing (ART/ADT)*."

ASAE Engineering Practice: ASAE EP456

SECOND DRAFT.
PREPARED BY Dr. LEV M. KLYATIS

TEST AND RELIABILITY GUIDELINES

Developed by the ASAE Testing and Reliability Committee

SECTION I—PURPOSE AND SCOPE

1.1 This Engineering Practice shows how product life can be specified in probabilistic terms, how life data should be analyzed, and presents the statistical realities of life testing, including a random rate.

1.2 The main purpose of the Engineering Practice is : evaluation of reliability measures and test procedures that give the information for this evaluation.

1.3 This Engineering Practice is not intended to be a comprehensive study on the large variety of subjects encompassed in the analysis of life testing, but rather a guide to demonstrate the power, usefulness and methodology of reliability analysis.

SECTION 2—BACKGROUND

2.1 A key characteristic of a product is its life expectancy. In the agricultural equipment industry a great deal of testing is conducted on materials, components and complete machines to determine expected life.

2.2 Product life needs to be specified in probabilistic terms.   This mandates, that test programs require, enough replications to minimize the risk of erroneous test conclusions. The test engineer and designer must establish a test program to balance test time and cost against risk of a product failure.

2.3 The primary emphasis is on the problem of quantifying reliability in product design and testing

2.4 The only way to measure system reliability is to test a completed product with a combination of influences or components, under conditions that simulate real life, until failure occurs. One simply cannot assess reliability without data, and the more data available, the more confidence one will have in the estimated reliability level. Extensive testing is often considered undesirable, because it results in expenditures of time and money. Thus, it must consider the trade off between the value of more confidence in estimated reliability versus the cost of more testing. It is easy to arrive at a test program's cost; the cost of not having the test program is difficult to calculate.

2.5 Before the reliability of a test subject is measured, certain procedural constraints must be established. For example, the amount of preventive maintenance permitted, if it is permitted at all, and the degree to which the system operator can participate in correcting failures must be specified, because the manner in which the system is operated can affect the calculated reliability level.

2.6 If the particular failure density and the distribution function are known, the reliability function can be found directly. The distribution of failure should consider different types of distribution (normal , Weibull , exponential , log normal , gamma , Student , Rayleigh , etc.) In practice, one is usually forced to select a distribution model without having enough data to actually verify its appropriateness. Therefore the procedures used to estimate the various reliability measures may be obtained from empirical data.

2.7 Type of Test for Reliability Testing and Planning: Functional Tests. Environmental Tests, Reliability Life Tests.

2.8 This Engineering Practice is used in to order to raise the level of knowledge and skill of test engineers and researchers who work in the area of testing and reliability, and increase the quality and productivity of machinery, thereby decreasing cost of design and test work.   This is especially crucial in the following areas: reliability measures and how to evaluate; environmental tests and the influence on the reliability; accelerated testing and evaluating of agricultural equipment; physical simulation of the life rates in the laboratory, and grounding of test regimes; testing not only details and units, but complete machines and equipment

2.9 The strategy of test for the evaluation of farm machinery reliability include: a) reception of initial information;b) simulation of operating conditions; c) accelerated testing; d) statistical applicants; e) reliability evaluation.

2.10 Testing and evaluation of reliability can be for different levels of machinery life: 1) for one work season; 2) for optimal product life; 3) for service life; 4) on the instruction of the buyers. The machinery can be tested in practice under ordinary operating conditions, laboratory conditions, and combinations of both these types of testing (Fig. 1.) Tests under laboratory conditions consist of (Fig. 1.) tests on the mechanical and environmental influences.

SECTION 3—TERMINOLOGY

3.1 Distribution: A mathematical function giving the cumulative probability that a random quantity, such as a component's life, will be less than or equal to any given value.

3.2 Probability: Likelihood of occurrence based on significant tests.

3.3 Sample: A small number of parts which will be considered as representative of the total population.

3.4 Accelerated test: A test in which the deterioration of the test subject is intentionally accelerated over that expected in service.

3.5 Accuracy: A generic concept of exactness related to the closeness of agreement between the average of one or more test results and an accepted reference value or the extent to which the readings of a measurement approach the true values of a single measured quantity.

3.6 Characteristic: A property of items in a sample population which, When measured, counted or otherwise observed, helps to distinguish between the items.

37 Control (evaluation): An evaluation to check, test, or verify.

3.8 Standard deviation: The most usual measure of the dispersions most Often to form an estimate of some population parameter.

3.9 Population: The totality of items or units of equipment under

1

Fig. 6.8   The first page of the second draft of the standard EP 456 "Test and Reliability Guidelines"

Drafts of these two standards, which were partly included in the author's book *Successful Prediction of Product Performance*, were published by SAE International in 2016 [4].

From the preparation of these standards and as a result of discussions on G-11 meetings, the author updated his theories on accelerated reliability and durability testing that had been developed earlier.

The testing methodology developed in these new standards was a key factor for prediction technologies of product reliability and other components of the product efficiency.

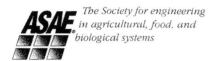

The Society for engineering in agricultural, food, and biological systems

April 25, 1996

Lev M. Klyatis
1290 E. 19th St., Apt. 5B
Brooklyn, NY 11230

Dear Lev Kylatis:

The China Society of Agricultural Engineers and the China Agricultural University have invited a delegation of professionals interested in agricultural engineering and technology to visit the People's Republic of
China in August. Your participation in this exclusive mission would contribute to its success, and as the
appointed delegation leader, I extend to you a personal invitation to join me and others on this exciting,
educational, and professionally rewarding project.

We will attend the International Conference on Agricultural and Biological Environment Engineering at the
China Agricultural University in Beijing. Other cities to be visited during our mission include Shanghai
and Hangzhou. Site visits are also planned at agricultural communes so that we can see, first hand, how
the Chinese are dealing with the latest technological changes in agriculture. A list of preliminary program
focal points has been enclosed for your review along with a day-by-day itinerary. There will also be
specially planned cultural and sightseeing programs for the delegation. Spouses and guests are invited and
encouraged to participate.

The delegation will depart for China on August 13, and return on August 24, 1996. The program, although
not sponsored by ASAE, is being coordinated by the American People Ambassador Program, a division of
the Citizen Ambassador Program of People to People International, for the professional, educational, and
cultural benefit of ASAE members. This study mission is designed to comply with Section 162 of the
Internal Revenue Code for potential deduction as a business expense; however, all participants should
consult with their tax advisor.

To enhance the quality of our professional programs, the size of the delegation is limited, and a timely
response is necessary to ensure an available place with the delegation. I ask that you contact Ms. Kathleen
S. Sieler, our Program Coordinator, as soon as possible if you have questions about this project. She can
be reached at (509) 534-0430, extension 422, in the Pacific time zone. We look forward to hearing from
you.

Sincerely,

Dr. Wayne Coates, International Director
American Society of Agricultural Engineers

ASAE ■ 2950 NILES ROAD, ST JOSEPH, MI 49085-9659 USA ■ VOICE 616.429.0300 ■ FAX 616.429.3852 ■ E-MAIL (internet) hq@asa

Fig. 6.9   Invitation Lev Klyatis to People's Republic of China as a member of American delegation

The presentations and publications of these standards were included in papers
for the SAE World Congresses in Detroit, journal articles, and other publications in
this area. They have proven to be valuable to everyone wanting to implement successful prediction of product efficiency.

As a result of the development of these standards and his career work in quality and reliability, the author was invited to work as a consultant and seminar instructor for several companies and organizations.

As a part of G-11 meetings, committee experts often visited companies in the geographic area near the meeting. Figure 4.12 in [3] shows G-11 members, including the author, during a visit to the NASA Langley Research Center.

These meetings helped the author better understand the methods in use at NASA and aerospace companies and of their understanding of how to obtain successful prediction of product efficiency. He was surprised to learn that it was difficult to find the person or group at the Langley Research Center who was responsible for reliability or durability testing. Ultimately it appeared their reliability/durability testing professionals were not directly involved in reliability and durability testing of the components or the completed product. Some of the documents related to these standardization efforts are presented below.

Figure 6.10 shows the author during a SAE G-11 Division meeting in Washington D.C. There he gave a presentation for the G-11 Division members that helped them better understand his concepts for successful efficiency prediction. An example was a discussion with Mr. Dan Fitzsimmons, The Boeing Company Technical Fellow, and Statistical QE Supplier Management & Procurement Commercial Airplanes, which helped him to understand how the bureaucratic structure of large companies make it difficult for them to implement the author's methodologies. Mr. Fitzsimmons learned how organizational structures often involve divisional hierarchies where each vice president had control and responsibility only for his or her direct area of

Fig. 6.10 Lev Klyatis during the SAE G-11 Division meeting (break time) in Washington DC, November 2012

responsibility. This practice violates what was noted above in that successful prediction of product efficiency requires these divisional interactions to be constant and integrated. Product success is not related to a single corporate area but must include all groups from basic areas of research, testing, design, production, management, and marketing. Only through this integration of effort can a company improve a product's life cycle cost, profit, and recalls. In Chap. 3, one can learn how the above relates to engineering culture improvement.

The author's experience and background as a test engineer and principal engineer at the state test centers in the USSR, Russia, and Ukraine, as well as his roles as chairman of State Enterprise TESTMASH, government of the USSR, university professor, consultant, laboratory chair, lecturer, and advisor for American industrial companies and universities helped him better understand how one can improve research, design, test, and manufacturing processes and technologies.

The lessons learned from these interactions were included in the author's work for American Society Agricultural Engineers (ASAE), American Society for Quality (ASQ), and SAE International G-11 and several presentations on efficiency prediction and reliability testing for the G-11 Division under the common title "Reliability Testing."

Presented below is an example of the meeting agenda with the presentation "Overview of the Role and Contents of Six G-11 JA1009 Reliability Testing Standards."

"SAE G11, Reliability, Maintainability/Supportability and Probabilistic Methods Systems Group Meeting Agenda

November 28-29, 2012. Washington, DC

Hosted by:

Honeywell International 101 Constitution Ave, Washington, DC 20001

Day 1: Wednesday–November 28, 2012

10:00-10:30   Coffee and greetings

10:30-10:45   Host welcome – *Honeywell (speaker: TBD)*

10:45-11:00   Agenda review

11:00-11:15   G-11 Division Overview: prior work, charter and discussion of go-forward plans - *Michael Gorelik*

11:15-11:30   SAE staff presentation –*Donna Lutz, SAE Aerospace*

11:30-12:00   Review of Sept. 2011 meeting minutes

12:00-1:00   Lunch Break

1:00 - 1:45   CBM Recommended Practices Guide Project status (Including RAMS-5 meeting and DoD Maintenance Symposium panel session debriefs) -- Chris*Sautter and JC Leverette*

1:45-2:30   Review of the status and plans by committee:

G-11Reliability
G-11Maintainability / Logistics
G-11Probabilistic Methods
Day 2: Thursday –November 29, 2012

8:30-8:45   Review of Day 2 Agenda

8:45-9:30   **"Overview of the Role and Contents of Six G-11 JA1009 Reliability Testing Standards" – *Lev Klyatis***

9:30-12:00   Committees breakout sessions (continued)

G-11Reliability
G-11Maintainability / Logistics
G-11Probabilistic Methods
12:00-1:00   Lunch Break

1:00-3:30   Committees breakout sessions (continued)

G-11Reliability
G-11Maintainability / Logistics
G-11Probabilistic Methods".

Following the presentation, the second standard was prepared, discussed, and approved for voting. Fortunately, this work resumed in 2017. Shown below is the SAE G-11 Spring 2017 meeting announcement.

SAE G-11 Reliability Maintenance Support and Probabilistic Methods Committees meeting (2017):.

G-11 Spring 2017 Meeting Notice.doc

| StandardsWorks@sae.org | All, I am please to announce that we will be having a G-11 Group... | Jan 6 (1 day ago) |

Documents to be Reviewed for Each Committee

G-11 Probabilistic Methods Committee

Document List

Document	Title	Date	Status
AIR5080	Integration of Probabilistic Methods into the Design Process	May 08, 2012	Reaffirmed
AIR5086	Perceptions and Limitations Inhibiting the Application of Probabilistic Methods	May 08, 2012	Reaffirmed
AIR5109	Applications of Probabilistic Methods	May 08, 2012	Reaffirmed
AIR5113	Legal Issues Associated with the Use of Probabilistic Design Methods	May 08, 2012	Reaffirmed

G-11R, Reliability Committee

Works in Progress

Project	Title	Sponsor	Date
ARP6204	Condition Based Maintenance (CBM) Recommened Practices	Fred Christian Sautter	Sep 15, 2011
JA1009	*Reliability Testing Standard*	*Lev Klyatis*	*Oct 27, 1998*
JA1009-1	*Reliability Testing - Strategy*	*Lev Klyatis*	*Apr 26, 2013*

G-11 Reliability Applications Committee

Document List

Document	Title	Date	Status
ARP5638	Rms Terms and Definitions	Mar 06, 2005	Issued

Meeting Location: **Rosen Plaza Hotel;** 9700 International Drive. Orlando FL, 32819

Meeting Time: 1.30pm – 4.00pm; Meeting Room: Salon 17

If you have any questions, please contact Sonal Khunti, Aerospace Standards Specialist, at +44
        In conjunction with RAMs Conference. Meeting: January 24th, 2017

7590184521 or sonal.khunti@sae.org."

Resulting from the discussion during the SAE International G-11 Division meetings, the following proposal concerning the JA 1009 Reliability Testing Standard was recommended (see the abstracts below).

Reliability Testing Standard JA 1009/A: Glossary (of Terms and Definitions)
Consists of a description of the applicable documents (SAE publications, ECSS publications, ASQ and Federal publications, IEC publications, US governmental publications, other publications, applicable references, and glossary of terms and definitions: accelerated reliability testing, accelerated durability testing, durability testing, reliability testing with appropriate notes, accurate simulation of field input influences, accurate prediction, accurate system of reliability prediction, an accurate physical simulation, classification accuracy, correlation, hazard analysis, hazard control, hazard reduction, human factors, human factors engineering, multi-environmental complex of field input influences, output variables, compliance test, laboratory test, accelerated testing, reliability, reliability improvement, reliability growth, durability, failure, failure mechanism, time to failure, time between failures, item, fault, system, subsystem, life cycle, useful life, characteristic, simulation, prediction, life cycle costing, special field testing, etc. (total of 11 pages)).

Reliability Testing Standard JA 1009/2: Procedures
Reliability testing (accelerated reliability/durability testing) as a key factor for accurate prediction quality, reliability, durability, maintainability, life cycle cost, and profit during given time and accelerated product development. Consists of a description of the step-by-step procedure for the design of reliability testing which is intended for any specific equipment to be tested when it is considered necessary to simulate closely the real conditions of use of the test subject. It applies fully to laboratory testing in combination with special field testing, etc.

Reliability Testing Standard JA 1009/1: R Reliability Testing—Strategy
Consists of basic concepts of reliability (accelerated reliability/durability) testing, accuracy of the simulation of field conditions (input influences in interconnection with safety and human factors), physical (chemical) degradation mechanism of the test subject as a criterion for an accurate simulation of field input influences, requirements to obtaining accurate initial information from the field, methodology for selecting a representative input region for accurate simulation of the field conditions, moving the field to the laboratory, etc. (total of 18 pages).

Reliability Testing Standard JA 1009/3: R Reliability Testing—Equipment
Consists of requirements for equipment for accelerated reliability testing and its components (multi-environmental, vibration, dynamometer, wind tunnel, etc.). Test equipment is considered as a combination of multi-environmental + mechanical + electrical (electronic) and other types of testing. Examples of equipment for reliability/durability testing in the laboratory, etc.

Reliability Testing Standard JA 1009/4: Statistical Criteria for a Comparison of Reliability Testing Results and Field Results

Consist of statistical criteria for comparison "during reliability testing" of the output variables and physics of degradation process with the output variables and degradation process in the actual service conditions, statistical criteria for comparison "after reliability testing" of the reliability indexes (time to failure, failure intensity, etc.) with these indexes in actual service, etc.

Reliability Testing Standard JA 1009/5: Collection, Calculation, Statistical Analysis of Reliability Testing Data, Development Recommendations for Improvement of Test Subject Reliability, Durability, and Maintainability

Consists of a methodology of data collection during the test time, statistical analysis of these data, methods for analysis of the reasons of degradation and failures, preparations and recommendations for eliminations of these reasons, counting the accelerated coefficient, and development recommendations for improving the reliability of test subject.

Beginning of draft of SAE International Standard Reliability Testing—Glossary can be seen below:

SAE *Internation* **AEROSPACE**	**SURFACE VEHICLE/ AEROSPACE RECOMMENDED PRACTICE**	**SAE** JA1009 PropDft XXX2012
		Issued Proposed Draft 2012-04-24
	R Reliability Testing--Glossary	

Rationale

JA1009, Reliability Testing, is being revised to broaden the technical content and to better organize the material.

Foreword

This standard defines words and terms most commonly used which are associated with Reliability Testing (accelerated reliability testing). It is intended to be used as a basis for reliability testing definitions and to reduce the possibility of conflicts, duplications, and incorrect interpretations either expressed or implied elsewhere in documentation.

1. Scope

This document applies to reliability testing that is performed to support aerospace applications.

1.1 Purpose

The purpose of this standard is to define words and terms used most frequently in specifying reliability testing (Accelerated Reliability and Durability Testing) to give these terms a common meaning for the contractors and users in aerospace.

2. References

2.1 Applicable Documents

2.1.1 SAE Publications

Available from SAE International, 400 Commonwealth Drive, Warrendale, PA 15096-0001, Tel: 877-606-7323 (inside USA and Canada) or 724-776-4970 (outside USA), www.sae.org.

ARP 5638	RMS Terms and Definitions

2.1.2 ECSS Publications

Available from European Cooperation for Space Standardization (ECSS).

ECSS Secretariat, ESA ESTEC, P.O. Box 299, 2200 AG Noordwijk, The Netherlands,
Phone: +31-71-565 5748
Fax: +31-71-565 6839, ecss-secretariat@esa.int

ECSS-Q-30B	European Cooperation for Space Standardization (ECSS) Space Product Assurance, Dependability
ECSS-Q-30B	Glossary of Terms

2.1.3 ISO and Federal Specifications

ISO 9000:2000	Quality Management Systems—Fundamentals and Vocabulary

2.1.4 IEC Publications

Available from International Electrotechnical Commission, 3, rue de Varembe, P.O. Box 131, 1211 Geneva 20, Switzerland, Tel: +44-22-919-02-11, www.iec.ch.

IEC 60050-191:1990	Quality Vocabulary—Part 3: Availability, Reliability, and Maintainability terms—Section 3.2, Glossary of international terms
IEC 60050-191 International Electrotechnical Vocabulary—Chap. 191: Dependability and Quality in Service	(see http://www.electropedia.org/iev/iev.nsf/index?openform&part=191)

2.1.5 U.S. Government Publications

Available from the Document Automation and Production Service (DAPS), Building 4/D, 700 Robbins Avenue, Philadelphia, PA.

19111-5094, Tel: 215-697-6257, http://assist. daps.dla.mil/quicksearch/MIL-STD-280	Definitions of Item Levels, Item Exchangeability, Models, and Related Terms
MIL-STD-721C	Definitions of Terms for Reliability and Maintainability
MIL-HDBK-781	Reliability Test Methods, Plans, and Environments for Engineering Development, Qualification, and Production
MIL-STD-882	System Safety Program Requirements

2.1.6 Other Publications

Chan, H. Antony, T. Paul Parker, Charles Felkins, Antony Oates, 2000, *Accelerated Stress Testing*. IEEE Press.

Klyatis Lev M., 2012, *Accelerated Reliability and Durability Testing Technology*, John Wiley & Sons, Inc.

Klyatis, Lev M., Eugene L. Klyatis, 2006, *Accelerated Quality and Reliability Solutions*; Elsevier, UK.

Nelson, Wayne, 1990, *Accelerated Testing*. John Wiley & Sons, New York, NY.

Reliability Toolkit, Commercial Practices Edition; Reliability Analysis Center, 1993.

Glossary of Terms and Definitions (*partly from the standard*)

Accelerated testing	Testing in which the deterioration of the test subject is accelerated.
Accelerated reliability testing or accelerated durability testing (or durability testing)	Testing in which:
	(a) The physics (or chemistry) of the degradation mechanism (or failure mechanism) is similar to this mechanism in the real-world using given criteria.
	(b) The measurement of reliability and durability indicators (time to failures, its degradation, service life, etc.) has a high correlation with these indicators measurement in the real world using given criteria.
Note 1	Accelerated reliability testing, or accelerated durability testing, or durability testing offers useful information for accurate prediction reliability and durability, because they are based on accurate simulation of real-world conditions.
Note 2	It is identical to reliability testing if reliability testing uses for accurate reliability and durability prediction during service life, warranty period, or other.
Note 3	Accelerated reliability and durability testing, such as accelerated testing, is connected with the stress process. Higher stress means a higher acceleration coefficient (ratio of time to failures in the field to time to failures during ART), lower correlation between field results and ART results, and less accurateness of prediction.

Note 4	Accelerated reliability and accelerated durability testing (durability testing) consists of a complex of laboratory testing and special field testing. The laboratory testing provides a simultaneous combination of a whole complex of multi-environmental testing, mechanical testing, electrical testing, and other types of real-world testing. The special field testing takes into account the factors which cannot be accurately simulated in the laboratory, such as the stability of the product's technological process, and how the operator's reliability influence on test subject's reliability and durability. Accurate simulation of field conditions requires full simulation of the field input influences integrated with safety and human factors.
Note 5	Accelerated reliability testing (ART) and accelerated durability testing (ADT) (or durability testing) have the same basis—accurate simulation of the field situation. The only difference is in the indices of these types of testing and the length of testing: reliability testing is usually the mean time to failures, time between failures, and other parameters of interest; for durability it is the amount of time or volume out of service.
Note 6	Accelerated reliability testing can be for different lengths of time, i.e., warranty period, 1 year, 2 years, service life, and others.

Prepared by SAE Subcommittee G-11R, Reliability of Committee G-11, Reliability, Maintainability, Supportability, and Probabilistic Methods

One can see below beginning of draft of SAE G-11, JA1009 standard Reliability Testing—Strategy

SAE *International*® AEROSPACE	**SURFACE VEHICLE/ AEROSPACE RECOMMENDED PRACTICE**	**SAE** JA1009/1 PropDft APR2013
		Issued Proposed Second Draft 2013-04-02
	Reliability Testing - Strategy	

Rationale

SAE JA1009, Reliability Testing, is being completely revised to incorporate technical content and to better organize the material.

Foreword

This standard is to provide a basis for obtaining the information for accurate prediction of reliability, quality, safety, supportability, maintainability, and life cycle cost in real-world conditions during any part of the life cycle (warranty period, service life, and others) and to reduce the possibility of conflicts, duplications, and incorrect interpretations either impressed or implied elsewhere in the literature.

The following criteria were used to determine whether to include content in this document:

1. Reliability testing is identical to accelerated reliability testing if reliability test-ing uses for accurate reliability and durability prediction during service life, war-ranty period, or other time. Therefore, the term accelerated reliability testing below is used often instead of term reliability testing.
2. The purpose of this SAE standard is to standardize the strategy that helps to obtain information for accurate prediction, during design and manufacturing, of reliability, durability, and maintainability in the real-world conditions.
3. The real-world conditions consist of full input influences integrated with safety and human factors.

Scope
This document describes reliability testing that is performed to support aerospace applications.

Purpose
The purpose of this standard is to define the strategy of accelerated reliability testing to give a common meaning of this strategy for contractors and users in aircraft, aerospace, and other areas.

References
Applicable Documents

The following publications form a part of this document to the extent specified herein. The latest issue of SAE publications shall apply. The applicable issue of other publications shall be the issue in effect on the date of the purchase order. In the event of conflict between the text of this document and references cited herein, the text of the document takes precedence. Nothing in this document, however, super-sedes applicable laws and regulations unless a specific exemption has been obtained.

SAE Publications
Available from SAE International, 400 Commonwealth Drive, Warrendale, PA 15096-0001, Tel: 877-606-7323 (inside USA and Canada) or 724-776-4970 (out-side USA), www.sae.org
SAE International Reliability Testing Standard JA 1009/1

ASTM Publications
Available from ASTM International, 100 Barr Harbor Drive, P.O. Box C700, West Conshohocken, PA 19428-2959, Tel: 610-832-9585, www.astm.org

ASTM E2696-09   Standard Practices for Life and Reliability Testing Based on the Exponential Distribution

ECSS Publications
Available from European Cooperation for Space Standardization (ECSS). ECSS Secretariat, ESA ESTEC, P.O. Box 299, 2200 AG Noordwijk, The Netherlands, Phone: +31-71-565 5748, Fax: +31-71-565 6839, ecss-secretariat@esa.int

ECSS-Q-30B	Space Product Assurance. Dependability.
ECSS-M-00-03A	Space Project Management. Risk Management.
ECSS-Q-40-02A	Space Product Assurance. Hazard Analysis.
ECSS-Q-40B	Space Product Assurance, Safety.

IEC Publications

Available from International Electrotechnical Commission, 3, rue de Varembe, P.O. Box 131, 1211 Geneva 20, Switzerland, Tel: +44-22-919-02-11, www.iec.ch.

CEI IEC 61508-1	Functional Safety of Electrical/Electronic/Programmable Electronic Safety-Related Systems. Part 1: General Requirements
IEC 60300–3–2, Ed.2	Dependability Management—Part 3-2: Application Guide—Collection of Dependability Data from the Field
IEC 60605-2	Reliability of Systems, Equipment, and Components—Part 10: Guide to Reliability Testing Section 10.2 Design of Test Cycles ISO Publications Available from American National Standards Institute, 25 West 43rd Street, New York, NY 10036-8002, Tel: 212-642-4900, http://www.ansi.org/
ISO 9000:2000	Quality Management Systems—Fundamentals and Vocabulary
ISO 14121	Safety of Machinery—Principles of Risk Assessment Joint IEC/ISO Publications
IEC/ISO Guide 51	Safety Aspects—Guidelines for their Inclusion in Standards
ISO/IEC JTC1/SC7	Software engineering—Software Life Cycle Processes—Risk Management

U.S. Government Publications

Governmental Publications

Available from the Document Automation and Production Service (DAPS), Building 4/D, 700 Robbins Avenue, Philadelphia, PA 19111-5094, Tel: 215-697-6257, http://assist.daps.dla.mil/quicksearch/

MIL-HDBK-108	Sampling Procedures and Tables for Life and Reliability Testing (Based on Exponential Distribution)
MIL-HDBK-217F	Reliability Prediction of Electronic Equipment
MIL-STD-690D	Failure Rate Sampling Plans and Procedures
MIL-STD-756B	Reliability Modeling and Prediction
MIL-HDBK-781	Reliability Test Methods, Plans and Environments for Engineering Development, Qualification, and Production
MIL-HDBK-781A	Handbook for Reliability Test Methods, Plans, and Environments, Engineering, Development, Qualification, and Production
MIL-STD-781D	Reliability Testing for Engineering Development, Qualification and Production
MIL-STD-882C	System Safety Program Requirements
MIL-STD-2074	Failure Classification for Reliability Testing
DoD 3235. 1H	Test and Evaluation of Systems Reliability, Availability, and Maintainability: A New Primer

Other Publications

Chan, H. Antony, T. Paul Parker, Charles Felkin, Antony Oates, 2000, *Accelerated Stress Testing*. IEEE Press.

Lev Klyatis, 2016, *Successful Prediction of Product Performance. quality, reliability, durability, safety, maintainability, life cycle cost, profit, and other components. SAE International.*

Klyatis Lev M., 2012, *Accelerated Reliability and Durability Testing Technology*, John Wiley & Sons, Inc.

Klyatis, Lev M., Eugene L. Klyatis, 2006, *Accelerated Quality and Reliability Solutions*, ELSEVIER. UK.

Nelson, Wayne, 1990, *Accelerated Testing*, John Wiley & Sons, New York, NY.

Reliability Toolkit: Commercial Practices Edition. Reliability Analysis Center. 1993.

Strategy of Accelerated Reliability Testing

Specific Components of Reliability Testing Strategy

The study of real-world data connection—conditions to determine the important parameters that are to be simulated in the laboratory.

Quantitative and Qualitative Data Collection

Quantitative data collection is the collection of data that can be stated as a numerical value. Qualitative data collection is the collection of softer information, for example, reasons for an event occurring. Both data types are important and support each other. The type of data collected will depend on the sort of questions to be answered by the data.

Contents and Methods of Data Collection

Data collection is providing for accurate simulation of real-world conditions for reliability testing. These conditions include full field input influences, output variables, safety and human factors, reliability, durability, and maintenance data. The conditions are collected over several years and involves many different users and maintenance personnel. Thus data collection is a large-scale effort with possible sources of data corruption. Accordingly, the data collection, collation, and recording process have to emphasize ease of use and error proofing.

Reliability, durability, and maintenance data are the result of field conditions action on the product and share many common elements. Therefore, reliability data collection should be integrated with the maintenance record system. Whenever possible, all data sharing of common elements should be integrated with the maintenance record system. The accuracy of data reporting can be increased if the reporting form is combined with other types of reporting, for example, economic compensation (spare costs, payment under guarantee, mileage compensation, and time reporting for the repair person). The quality of the reporting improves if the repair person knows how the data will be used. Additionally, they should be notified if their data reporting is incomplete or ambiguous.

Data collection can be automated or semiautomated by using electronic data logging devices. The most complex data collection uses built-in electronics to perform the same task.

Automated data collection (ADC), also known as automatic identification (AutoID) and automatic identification and data capture (AIDC) (incorrectly referred to as "bar coding" by many), consists of many technologies including some that have nothing to do with bar codes.

Field Conditions

There are three integrated complexes present in the field conditions:

- Full complex of real-world input influences
- Safety problems
- Human factors

The contents of the first complex as well as two others are common for many types or similar groups of products and processes. The details are more specific for each application. For most of them, the full field input influences consist of multi-environmental, mechanical, electrical, and electronic groups. Each group is also a complex of subcomponents. A multi-environmental complex of field input influences consists of temperature, humidity, pollution, radiation, wind, snow, fluctuation, air pressure, and rain. Some basic input influences combine to form a multifaceted complex. For example, chemical pollution and mechanical pollution combine in the pollution complex. The mechanical group of input influences consists of different less complicated components. The specifics of this group of influences depend upon the product or process details and functions. The electrical group of input influences also consists of several diverse types of simpler influences such as input voltage, electrostatic discharge, and others. These factors are interdependent, interconnected, and interact simultaneously in combination with each other. Accurate simulation needs simulation of the above interconnection.

Safety problems are the combination of two basic components: risk problems and hazard analysis. Both are interconnected with reliability. For example, failure tolerance is one of the basic safety requirements that are used to control hazards. Another example is safety risks that are the result of the hazardous effects of failure functions. Fault tree analysis may be used to establish a systematic link between the system-level hazard and the contributing hazardous event at the subsystem, equipment or piece-part level.

The same logic applies to the solution of risk problems. There are many standards in reliability and safety that include the interconnection of reliability with safety.

Each of two basic safety components is a combination of subcomponents.

The solution to the risk problem is found in the following subcomponents:

- Risk assessment
- Risk management
- Risk evaluation

Each of the above subcomponents consists of sub-subcomponents.

Solving the safety problem requires the simultaneous study and evaluation of the full complex of these interacting components and subcomponents.

To obtain information for risk assessment, one needs to know the following:

- Limits of the machinery
- Accident and incident history
- Requirements for the life phases of the machinery
- Basic design drawings that demonstrate the nature of the machinery
- Statements about damage to health

For risk analysis, one needs the following:

- Identification of hazards
- Methods of setting limits for the machinery
- Risk estimation

Human factors engineering is the scientific discipline dedicated to improving the human-machine interface and human performance through the application of knowledge of human capabilities, strengths, weaknesses, and characteristics.

Human factors always interact with reliability and safety because the reliability of the product has a connection with the operator's reliability and capability. It is called human factors in the USA, whereas in Europe, it is often called ergonomics.

This is an umbrella term for several areas of research that include as follows:

- Human performance
- Technology
- Human-computer interaction

The term used to describe the interaction between individuals and the facilities and equipment and the management systems is human factors. The discipline of human factors seeks to optimize the relationship between technology and humans. The human factors apply information about human characteristics, limitations, perceptions, abilities, and behavior to the design and improvement of objects and facilities used by people. The essential goal of human factors is to analyze how people are likely to utilize a product or process.

It is necessary to provide special field testing for the evaluation/prediction of the stability of a product/technology function for the product and how the management and operator's reliability influence the product's reliability (Figs. 6.11 and 6.12).

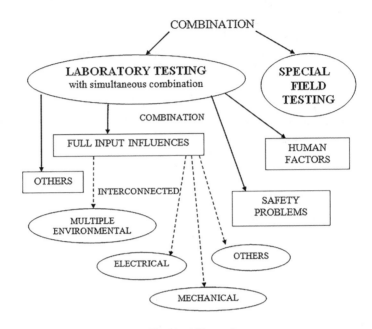

Fig. 6.11 Scheme of accelerated reliability/durability testing

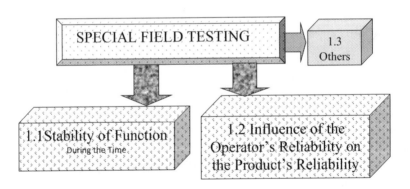

Fig. 6.12 Scheme of special field testing
Prepared by SAE Subcommittee G-11R, Reliability of Committee G-11, Reliability, Maintainability, Supportability, and Probabilistic Methods
The author made the oral/written presentations about above standards at the SAE World Congresses (see [11, 12]).

6.4.3   Implementation of Reliability Testing During the Work for the International Electrotechnical Commission (IEC), US Representative for International Organizations in Standardization (ISO), IEC/ISO Joint Study Group in Safety Aspects of Risk Assessment

Following his presentation at the ASQ Quality Congress, 1999, the author received an invitation from Mr. John Miller, Chair of US Technical Advisory Group for Technical Committee 56, International Electrotechnical Commission (IEC) for international standardization, to be included in the US Technical Advisory Group—the group of experts representing the USA at the IEC. Accepting this invitation, he sat in the Technical Committee TC 56 Reliability and Maintainability. Figures 6.13, 6.14, 6.15 and 6.16 show the author's work in this committee and in the IEC International Congress. The Technical Committee TC 56 Dependability covers this area for all IEC committees in specific technical areas of energy, electronic, electrical, nuclear, and others.

The following is one story from the IEC International Conference in Beijing (China), where the author received an award from China, as can be seen in Fig. 6.16. A group of this meeting's members visited the Chinese National Institute, which was involved in the testing of electrotechnical and electronic devices. At this time, Dr. Klyatis saw the opportunity for improving their reliability testing of electronic and electrical devices in the area of practical reliability and durability testing technology.

During the committee's work, Mr. Valter Loll, from Nokia, Denmark, chairman of the TC 56 Committee, proposed that Lev Klyatis prepare a new IEC standard(s) in accelerated reliability testing. Dr. Klyatis subsequently prepared a first draft of this new IEC standard *Equipment Reliability Testing. Accelerated Testing of Actual Product.* But during this time, Mr. John Miller retired as chair of US Technical Advisory Group and this work was stopped. Later the ideas and information of this draft standard were used as the basis for preparing a group of six SAE International standards under the common title "Reliability Testing" (see above).

One can see in Fig. 4.15 [3] photo with the author during meeting IEC TC 56 Committee.

Unfortunately, because US representatives did not receive compensation for participating at the meeting and due to schedule and workload conflict, the IEC standards were not always advanced in a timely manner. These issues are common problems that hinder the advancement of other US and international standards.

Over 100 IEC technical committee members attended the SAE congress in Beijing. During the congress the majority of the work was through the plenary meetings, with the technical committee's work being performed separately.

A major part of the IEC technical committees work was reviewing proposals from countries for accepting their standards as recognized international. These committees were also responsible for the development new standards.

There are other organizations in the world that are developing international standardization. For example, the International Standardization Organization (ISO) is one such widely known international standard organization. Together they are planning and developing worldwide standards for all types of equipment.

For coordinating the methodological aspects for advancing standardization, IEC and ISO have joint study groups which shared areas of interest. The author worked as an IEC representative at one such joint study group, the IEC/ISO Joint Study Group in Safety Aspects of Risk Assessment. In Figs. 6.17 and 6.18 below are the

Yo

 United States National Committee of the International Electrotechnical Commission

A Committee of the American National Standards Institute

11 West 42nd Street   •   New York, NY 10036   •   (212)642-4936

FAX: (212)730-1346
(212)302-1286 (Sales Only)

8 September 2000

Dr. Lev Klyatis
ECCOL Inc.
225 Broadway, Suite 1410
New York, NY 10007

Subject: Delegate Accreditation Letter

Dear Dr. Klyatis:

The U.S. National Committee of the IEC is pleased to confirm your appointment to the USNC delegation for the announced IEC/TC 56 meeting. An Accreditation and Identification Card is enclosed for your use.

As you know, at this meeting you will represent the USNC/IEC. The positions on the technical agenda items will have been determined by the US Technical Advisory Group. Positions on policy and administrative matters should be developed in consultation with the USNC office. This will assure a unified US position at all levels of the IEC organization. For your information also please find the website link to the booklet "Guide for U.S. Delegates to IEC/ISO Meetings."
http://web.ansi.org/public/library/intl_act/default.htm

The USNC is grateful to you and your employer for agreeing to support the voluntary standards system and for your willingness to present US positions to IEC.

Sincerely,

Charles T. Zegers

Charles T. Zegers
Secretary, USNC/IEC

CTZ:dn
Encl.

Copy to: J.A. Miller
N.H. Criscimagna
J. Rennie

Fig. 6.13 The letter from the general secretary of US National Committee for IEC about accreditation Lev Klyatis as US representative for IEC

INTERNATIONAL ELECTROTECHNICAL COMMISSION TECHNICAL COMMITTEE No. 56 DEPENDABILITY Working Group 2: Methodology	56/WG2/PT 2.8 January 2001

Project 2.8: Revision of IEC 60300-3-1

Date: January. 20, 2001 in Philadelphia

Location   Philadelphia Marriott
             1201 Market Street
             Philadelphia, PA 19107 , USA
             Phone: +1-215-625-2900
             Fax: +1-215-625-6101
             Room: 406 (10:00 – 17:00)

Preliminary agenda

1   Opening of the meeting

2   Approval of the agenda

3   Current membership

Prof. Bobbio, Andrea	Italy	T. +39 01 1670 6742	bobbio@di.unito.it
Dr. Braband, Jens	Germany	T. +49 531 226 3231	Jens.Braband@vt.siemens.de
Mr. Jones, Jeffrey A.	United Kingdom	T. +44 / 1203 572 616	jones_j@wmgmail.wmg.warwick.ac.uk
Mr. Kleinrath, Peter	Austria	T. +43 5 1707 35918	peter-viktor.kleinrath@siemens.at
Dr. Klyatis, Lev	USA	T.	lev-nellya-klyatis-1-assoc@worldnet.att.net
Ms. Krasich, Milena	USA	T. +1 508 766 1134	Milena_Krasich@bose.com
Mr. Lohmar, John William	USA	T. +1	lohmarjw@navsea.navy.mil
Dr. Masuda, Akihiko	Japan	T. +81 554 63 6956	amasuda@ntu.ac.jp
Mr. Midgette, William H.	USA	T. +1 301 443 2536 ext 12	whm@cdrh.fda.gov
Mr. Miller, John A.	USA	T. +1 714 842 4776	millerja@earthlink.net
Mr. Sateesh. Kris	USA	T. +1 / 941 739 4229	kris.sateesh@edwards.spx.com
Mr. Schwarz, Erich	Germany	T. +49 89 636 48696	erich.schwarz@mchp.siemens.de
Dr. Turconi, Giorgio	Italy	T. +39 02 4388 8195	giorgio.turconi@italtel.it
Dr. Walls, Lesley	United Kingdom	T. +44 141 548 3616	lesley@mansci.strath.ac.uk
Dr. Wild, Antonin	Canada K2H 7X7	T. +1 613 820 0410	tonywild@echelon.ca
Mr. Loll, Valter	Denmark	T. +45 20 73 01 15	Valter.Loll@nmp.nokia.com

4   Confirmation of minutes 56(WG2/PT2.8)03 (LA meeting)

5   Further work on the revision of IEC 60300-3-1

Document "Revision IEC 60300-3-1 Draft Jan-2000 c.doc"
considering the comments in 56(WG2/PT 2.8)06 and 56(WG2/PT2.8)05

6   Time schedule for the revision, allocation of work

7   Date and place of the next meeting

8   Close of the meeting

- 1 -

Fig. 6.14   Preliminary agenda of IEC/T-56 Committee meeting

Fig. 6.15 Lev Klyatis expert on the US Technical Advisory Group for International Electrotechnical Commission (IEC) in Sydney (Australia) during an IEC meeting

Fig. 6.16 Dr. Lev Klyatis receiving award from China (Beijing) during the IEC Congress

documents from this group meeting in Frankfurt am Main (Germany), 2004; Fig. 6.17 highlights analysis prepared by the author that was considered during this meeting in Germany.

The author saw the need for improving the level of testing electronic and electrical devices, especially in practical reliability and durability testing, and why the adoption of his ideas in IEC and ISO standards was so important.

6.5   Implementation Through Author's and His Colleagues Seminars, Publications, Lectures, and Presentations over the World

The basic challenge of implementing any novel approaches is the innovative technology must, first, be embraced and accepted in the minds of the people who will be doing the implementation. This is regardless as to whether this implementation be engineering new methods or devices in research, design, or specific industrial company applications. Only when the professionals who will be implementing the new original approaches fully understand and are committed to the modern technology will successful implementation be possible.

The author's strategy, approaches, and innovations to successful efficiency prediction and accelerated reliability/durability testing (see Chaps. 3 and 4) were learned from his publications and accepted by others and then implemented in different areas of engineering, physical, and mathematical sciences and practices.

An important aspect of successful implementation of latest ideas and technologies is not only how new or high quality is the idea or technology and its benefits but also how expensive it will be to implement them.

This includes not just the direct costs of these technologies in comparison with present technologies but how their implementation will affect the costs (or savings) throughout the product's life cycle.

Consider the following example of successful accelerated reliability/durability testing. For this implementation it was important to do a cost vs. benefit analysis for the proposed accelerated reliability testing. The new test plan entailed a simultaneous combination of different types of input influences, as compared to separate simulations of each input influence. What would be the respective costs for the two methods was the question.

In this case, the costs of the equipment, the methodology, and conducting each of the single tests (e.g., separate vibration testing, or corrosion testing, or temperature plus humidity testing, or testing in dust chambers, or input voltage plus vibration testing) may appear to be less costly as compared to reliability testing using the simultaneous combination of simulation of all the above types of influences.

While this may seem correct, it only compares the direct cost of the two approaches to testing and is not sufficient to cover all of the affected variables and their associated costs. The quality of the testing level influences the cost of many

ACOS/JSG-ISO/Klyatis/4

SAFETY ASPECTS OF RISK ASSESSMENT

September 8, Frankfurt am Main

SUMMARY

1. Assessment of Current Publications in the Area

It was reviewed Standards of 3 international organizations were reviewed: IEC, ISO, and ECSS (European Cooperation for Space Standardization).

A.*ISO – 1 standard*: ISO 14121 "Safety of Machinery. Principles of Risk Assessment". This was prepared by the European Committee for Standardization (SEN) (as EN 1050:1996) and was adopted by Technical Committee ISO/TC 199 "Safety of Machinery". This consists of 18 pages including title page, foreword, introduction, contents, scope, normative references, terms & definitions, bibliography, one empty page. So, the contents of this standard cover 13 pages, including 6 pages of Annex A (informative) and Annex B (informative).

Assessment results:
1. This standard is for machinery only which is one of many aspects of safety risk. The biological, medical, chemical, food and other aspects are not considered.
2. It includes a small (but not main) fraction of methods for analyzing hazards and estimating risk.
3. Similarly the title directions sometimes do not correspond to IEC and ECSS standards. For example, in the Simulation area in this standard only "…B.5 Fault simulation for control system" which is not essential for safety risk assessment. No less important is the "simulation of dynamic of input influences during service life", "simulation of stress conditions (vibration, environmental conditions, operator's reliability)", and others that are in IEC standards. In ECSS standards there are "dependability analysis", etc.
4. No traffic-crash aspects of risk safety are included.
5. There is no coordination with standards of other standartization organizations.
6. Other possibility may occur.

B. *ECSS* - Assessed 7 standards: 1.Glossary of Terms, 2.Space Project Management (Risk management), Space Product Assurance (3.Safety, 4.Hazard Analysis, 5.Quality Assurance for Test Centers, 6.Dependability, 7. Software Product Assurance).

Assessment results:
1. Do not take into account current 75 IEC standards in Basic Safety and 57 IEC standards in Dependability.
2. The standard in Dependability (in comparison with 57 IEC standards in Dependability) consists of nothing about techniques of dependability and no examples.
3. In standard Quality Assurance for Test Centers there are no examples.
4. Simularly by the title directions do not correspond to the contents to IEC and ISO standards. See above example (A3).
5. Standards are not coordinated with standards of other standardization organizations

(IEC).

2/3

Fig. 6.17 The proposal by Dr. Lev Klyatis, an expert of ISO/IEC Joint Study Group in Safety Aspects of Risk Assessment, analysis of international standards in safety aspects of risk assessment. After that he proposed the way for their improvement

	ACOS/JSG-ISO/Sec/7
	2004-10-01

INTERNATIONAL ELECTROTECHNICAL COMMISSION

ADVISORY COMMITTEE ON SAFETY (ACOS)
Joint Study Group with ISO – Safety Aspects of Risk Assessment

SUBJECT

Report of the Joint Study Group with ISO meeting held in Frankfurt, Germany on 2004-09-08

1. Participation

Mr. F. Harless	JSG Convenor, IEC TC 44
Mr. G. Alstead	IEC TC 56
Mr. R. Bell	IEC SC 65A
Mr. N. Bischof	IEC TC 62 and SC 62B
Mr. E. Courtin	IEC TC 62
Mr. V. Gasse	IEC TC 108
Mr. H. Huhle	ACOS
Mr. L. Kylatis	IEC TC 56
Mr. I. Rolle	DKE
Mr. S. Rudnik	IEC TC 44
Mr. Y. Sato	IEC TC 56 and SC 65A
Mr. H. von Krosigk	IEC SC 65A
Mr. D. Cloutier	ISO TC 199
Mr. R. David	ISO TC 199
Ms. N. Stacey	ISO TC 199
Mr. M. Casson	Secretary, IEC CO

2. Opening of the meeting – introduction of the delegates
Mr. Harless, the JSG Convenor opened the meeting stressing that it was a Joint Study Group with ISO participation. He noted that there were three delegates from ISO/ TC 199 and that Mr. Courtin also represented ISO/ TC 210. The Secretary was requested to ensure that the report of the meeting is circulated within ISO.

The delegates then introduced themselves giving brief information on their professional activities and their IEC/ ISO affiliations.

Mr. Rolle welcomed the delegates to DKE.

3. Approval of the agenda
Mr. Harless proposed to kick-off the meeting with a presentation and then a round-table discussion should start prior to the lunch break.

The agenda was approved.

4. Presentation – Mr. Harless
Document: ACOS/JSG-ISO/Harless/3

Fig. 6.18 Document demonstrating Lev Klyatis as an Expert of ISO/IEC Joint Study Group in Safety Aspects of Risk Assessment

subsequent processes, including design, manufacturing, usage, warranties and recall, and service life. For example, the confidence levels of different qualities of testing affect return and recall rates, future changes in safety, quality, and reliability problems during the product's life cycle. This is a clear case where not accurately simulating the interactive real-world influences would lead to increased overall expenses and decreased economic profitability from the product.

It is the author's experience from consulting with many companies that when a test engineer working with vibration testing was asked what he was evaluating, usually, the answer was "reliability." The author would then explain why this answer is not truly correct. The product's reliability depends not only on vibration. Vibration is only one component of the many influences on product experiences in real world use. Such mechanical influences act in combination with multi-environmental influences, electrical and electronic influences as shown in this book. Together these product's degradation processes include time to failures, mean time between failures, and human factors.

If one does not take all of these into account, one cannot successfully predict the product's reliability or other components of efficiency.

As a result of testing with inaccurate field conditions simulation, unpredicted accidents, product failures, and other faults that were not identified in reliability testing may result in increased and unanticipated costs during usage. The incremental costs incurred during usage cycles and during the design and manufacturing testing omissions are often overlooked when estimating the cost attributed to testing. Such unseen costs invariably dwarf the savings gained by omitting needed testing.

As shown before in this book, over a 19 years' period from 2000 to 2019 recalls alone in the automotive industry have increased dramatically and led to industry losses of billions of dollars.

For these reasons, accelerated reliability (or accelerated durability) testing is actually more economical than independent single influence testing. While the testing may be more expensive, it yields decreasing costs during the manufacturing and the product life cycle. While this is true, unfortunately too often it is difficult to quantify these post-testing savings. Detailed consideration of this problem with examples is presented in [1, 2, 5].

As has been mentioned, the author's papers about the essence of his development of ART/ADT were discussed and published (AGRI/MECH/43) by the United Nations Economic Commission for Europe. The United Nations also published in New York his final paper *Accelerated Testing of Agricultural Machinery*. These papers presented clearly to the world the costs and improved benefits occurred by his testing methods.

The government of the Soviet Union first organized an engineering center that implemented the author's research results. Next organized was the State Enterprise TESTMASH for widespread implementation by other companies around the country. They utilized his ideas, research results, and innovations in reliability/durability testing and successful efficiency prediction.

The author was also the Soviet Union representative to the US-USSR Trade and Economic Council, which was organized by the US and the USSR governments for improving collaboration between both nations. At a meeting of this council in New York, there were two presentations from each country. The author made one of these, a presentation about the TESTMASH's solutions for development and implementation ART/ADT and efficiency prediction.

Following this presentation many executives of American companies meet with the author. They were anxious to discuss his approaches for improving reliability/durability testing and prediction using modern technologies. Two companies, including Steptoe and Johnson Company, were especially interested in implementing TESTMASH's approach into their operations. One result of the author's presentation was a contract by them with KAMAZ, Inc., as described above in Chap. 6.

Following the widespread implementation of TESTMASH, *All-USSR Journal Tractors and Farm Machinery* [8] published an interview with the author in which he discussed why he had been appointed by the USSR government as the chairman of this organization, how this was the new state-of-the-art research center for industry, what kind of problems this organization was designed to solve, and the role of TESTMASH in the development of the Soviet Union industry [8]. The basic ideas presented in this interview, the concepts, and the developed approaches are now realities which can be sustained well into the future. This interview also described many topics. Among them were the new strategy and methodology and specific test equipment being developed by TESTMASH, and in addition he also described a new industry standard for testing in the Soviet Union for the future 10 years. Elsewhere, it continues to be used in many countries throughout the world.

This was in stark contrast with author's experience with the management of many companies after coming to the USA.

Following presentations at annual SAE World Congresses from 2000 until 2022, there were other meetings with attending professionals who understood how useful this new direction in successful prediction of product efficiency could be.

For example, Mr. Richard Rudy, senior manager in quality and testing DaimlerChrysler, commented on how he believed this original approach was very interesting and could be useful for other industries. Further, he said he would be happy to give excellent recommendations or reviews of these ideas. Mr. Richard Rudy was in a unique position to carry out his promises. For many years he was the SAE International representative in the Annual Reliability and Maintainability Symposium (RAMS) Board of Directors and executive committee for a group of technical sessions under the common title "Integrated Design and Manufacturing (IDM)" and the annual SAE World Congresses in Detroit.

True to his word, Mr. Rudy wrote published positive reviews to author's books (see below). The author began consolidating his ideas, strategy, and implementation approach for efficiency prediction and reliability testing in the USA from presentations at the ASAE (American Society Agricultural Engineers) international meetings, RAMS, and ASQ Quality Congresses. In order to obtain money for trips to these meetings, he began to work as a fish delivery (Fig. 6.19).

Fig. 6.19 Professor Klyatis first job in the USA—delivering fish

When Elsevier published his first English book *Accelerated Quality and Reliability Solution* in the UK, Richard Rudy wrote a book review, which was published in the journal of *Total Quality Management and Business Excellence* (UK) in September 2006 (Fig. 6.31). When they met a few years later during another SAE World Congress in Detroit, Mr. Rudy described how after their first meeting he suggested to DaimlerChrysler's management team that they hire the author as a consultant, as it would help the company to improve their product reliability. Unfortunately, he was told there were no vacant positions or funding to retain the author.

Richard Rudy also wrote a review to author's book *Accelerated Reliability and Durability Testing Technology*, which was published by Wiley in 2012.

From beginning his life in America in 1993, the author quickly understood that many professionals involved in the fields of prediction in engineering, especially efficiency and reliability prediction and reliability testing, were not ready to learn and use the entire methodology of successful prediction of product efficiency. Therefore, he developed a method of implementing his solutions using a step-by-step approach. This approach presented in his first book in English *Successful Accelerated Testing* (Mir Collection, New York, 2002) that includes a basic step—the need for accurate simulation of field conditions for accelerated reliability testing. This method is easily understood by most people.

This is why, for most of his early years in the USA, the author focused his presentations on the need for accurate simulation of field conditions. He emphasized how this starting point was important, because so many industrial companies research centers were not accurately simulated the field conditions, and, as a result, laboratory testing results often were and continue even today to be different from the field (real-world) results.

The author's second book in English, which was much more complicated, demonstrated solutions to actual reliability and quality problems. After making his presentation at the SAE.

World Congress in Detroit which once again was attended by many people, one attendee came to the author and asked: "do you have experience in book writing?" The answer was—"Yes!," which was quickly followed by the question: "Would you be interested in preparing the book on your theories and strategy on quality and reliability for publishing by Elsevier?"

The author knew that Elsevier was a large world-class publisher and the person who had asked him was Jonathan Simpson, their commissioning editor, Science and Technology Books, Elsevier. He explained the process, whereby once he received a completed book proposal in the UK, they would send this book proposal for reviews, and, if the reviews were positive, they would sign an agreement with him for publishing the book. Elsevier received excellent reviews to the book proposal, and the *Accelerated Quality and Reliability Solutions* was published in 2006 [2]. This book is now available in many libraries throughout the world, and there are many published citations and reviews about it.

The story leading to the author's third book since moving to America also has an interesting background.

During the questions and answers segment, of the author's presentation at the SAE 2008 World Congress in Detroit, one attendee responded: "I work in durability testing. I have looked for literature on how I can conduct this testing, but I found only papers, journal articles, and books, where all the presented information is about durability testing, but no information of how I can conduct durability or reliability testing. I need information on the technology for doing this testing. What can you recommend?"

The author replied: "You are right. Presently there is not much published on how to do this. So, as soon I can prepare a new book in reliability and durability testing technology. It will be a guide for you and other professionals on how one can practically provide accelerated reliability and durability testing technology. I will let you know. Please, wait for my new book."

While the answer was impromptu, this question and the need for a publication on this subject motivated the author to write this third book in English. He understood that this book addressed problems needing solutions that would be applicable worldwide.

This required a publisher who could attract a worldwide audience. Years earlier while in Manhattan, New York, the author remembered seeing a building with large letters Wiley and Sons, and for many years he had dreamed of someday authoring a book that would be published by them. Therefore, he decided to send a book proposal first to Wiley to see if they would be interested in publishing it.

Wiley published this author's book *Accelerated Reliability and Durability Testing Technology* in 2012. While this proved to be a more difficult book to write, it provided a useful tool for testing professionals and it became very popular in industry as can be seen through an Internet search. Such a search does not include all references, but one can see that professionals are using his book (see [5]) in many different areas of industry and science worldwide.

The following examples are positive reviews of the author's six books in 2021:

"52 Best Automotive Engineering Books of All Time

As featured on CNN, Forbes and Inc – BookAuthority identifies and rates the best books in the world, based on recommendations by thought leaders and experts.

The #12 in this list has the following book:

Trends in Development of Accelerated Testing for Automotive and Aerospace Engineering

Lev M. Klyatis PhD Habilitated Dr.-Ing. Sc.D.

2020

Rate 4.16.

Accelerated testing (most types of laboratory testing, proving ground testing, intensive field/flight testing, any experimental research) is increasingly a key component for predicting of product's/process performance.

Congratulations!

Your book "Trends in Development of Accelerated Testing for Automotive and Aerospace Engineering" made it to BookAuthority's best <u>Automotive</u> Engineering books!

Trends in Development of Accelerated Testing for Automotive and Aerospace Engineering
By Lev M. Klyatis PhD Habilitated Dr.-Ing. Sc.D.

Trends in Development of Accelerated Testing for Automotive and Aerospace Engineering
Lev M. Klyatis PhD Habilitated Dr.-Ing. Sc.D.
|2020|

4.16 , #41

https://bookauthority.org/books/best-automotive-engineering-book

The books [1–6] reflect a step-by-step strategy and practical technology for providing advanced reliability/durability testing for successful prediction. While previous books were primarily descriptions of the author's basic work in this new direction of successful prediction of product efficiency, these references are very detailed and useful for immediate use on current problems and programs.

One example of the practical application of these theories involved author's work with Nissan. Having attended the presentation "Nissan speeds up truck bed durability testing" by Nissan Technical Center representatives at the SAE 2008 World Congress, Mr. Takashi Shibayama, the vice president of Jatko Ltd. (Japan) asked the author to conduct a seminar with the engineering directors of Nissan (Figs. 6.20 and 6.21) for improving their product quality and reliability. The author quickly agreed and during 2013 explained that his examples were a powerful demonstration of

Dear Dr.Lev Klyatis

From Nissan, 2 persons are going to attend the meeting.

Mr. Kaoru Onogawa
    Expert Leader , Powertrain Engineering Division

Mr. Shigao Murata
    Expert Leader , Powertrain Engineering Division

Expert leaders are the most high position as engineers in Nissan.
They report to directly to board Member , repot to Fellow (Similar to CTO
,Chief Technology Officer))

Mr Onogawa has grown in Engine Department, and Mr. Murata has grown
in Drive train Department.

Best Regards

Takashi Shibayama

Fig. 6.20 The letter of Mr. Takashi Shibayama, vice president of Jatko Ltd. (Japan) to Dr.
L. Klyatis about seminar meeting for two expert leaders of Nissan

ineffective coordination among different departments of the company and that supply chain management changes were needed to integrate corporate activities among
all of their directorates if they wished to improve reliability and durability testing
and efficiency prediction. It should be noted that this problem was not unique to
Nissan but can be found in many other small, midsized, and large industrial
companies.

SAE International has published over 20 of the author's papers at the SAE World
Congresses in Detroit. One of his presentations "Analysis of Different Approaches
to Reliability Testing" was provided at the congress that celebrated 100 years of
SAE (Fig. 6.22).

The RMS Partnership, a US Department of Defense (DoD) oriented organization, has continually promoted the implementation of reliability testing and efficiency prediction. The partnership highlighted this author's tutorials and
presentations in organized meetings and published his articles in their "Journal of
Reliability, Maintainability, and Systems Engineering."

They have also published reviews of his books. The RMS Partnership has also
organized some conferences with the invitation of the author as a speaker. For
example, Fig. 6.23 shows part of a program RMS Conference in San Diego (CA) in
a Systems Engineering Environment AQ Governmental-Industry Conference in
DAU West, San Diego, CA.

Information for the meeting with Nissan and Jatco Ltd. representatives.

A. ART/ADT implementation for accurate prediction of product quality, reliability, durability, safety, and human factors needs System of Systems approach which can be successful with using only interdisciplinary team. This is long term work for planning our collaboration. We have team of high level experts in advanced testing, prediction, medical, software, physical, design, and others who can provide consultant work for any industrial company.

B. Regarding Nissan's representatives questions:

1. Short answer see in slides 31 and 34 of "Chat with the Expert" (Keynote Speaker Dr. Lev Klyatis) at the SAE 2012 World Congress. Wide answer sees in Dr. Klyatis's whole Chat... and books published by Elsevier (Oxford, UK, 2006) and Wiley (US, 2012). These books include overview of 241 references over the word in the first book and 257 references in the second book
The above slides can be described in more detail during our meeting April 25, 2012.

2. This question relates to our expert who is Dr. of Medical Science, professor who works together with Dr. Lev Klyatis. He said that any specific problem solution in human factors and safety can be considered during detail program of our consultant work for implementation for Nissan of ART/ADT as a key factor for accurate prediction of product quality, reliability, safety, and durability.

3. Correlation of failure mode and failure mechanism is old criteria that cannot be useful for accurate prediction of the product life cycle cost, quality, reliability, and durability. One can see the results of use these criteria, for example, in a lot of recalls (relates to Nissan also) that leads to higher cost, lower safety, lost time than was predicted during design and manufacturing. For more detail understanding this problem, one needs to study real problems in automotive industry (and not this industry only) of inaccurate prediction, read carefully both mentioned Dr. Lev Klyatis's books, as well as his other dozens publications.
We can show, as examples, that many companies want to save money and time for product testing, therefore use simple vibration testing (sometimes call them "durability testing"), without calculation how much money will be loss for subsequent processes after this testing. We can show also, as examples, that many companies use vibration testing for mobile product in one degree of freedom, instead of six degrees of freedom as in real world, then loss a lot of money for improvement the low quality of their product. One more example from vibration testing: one uses sinusoidal load inputs instead of random load inputs as in real field. It is cheaper, but one does not want to understand that the companies will loss much more money for subsequent procedures. The recall is one from basic result of inaccurate simulation of real life for testing during design and manufacturing, inaccurate evaluation then prediction. This is only one from many problems from poor testing.
Finally, company has several times less profit that was predicted during design and manufacturing (see examples about above in Dr. Klyatis's books).

Fig. 6.21 The first page of Dr. Lev Klyatis' information for the meeting with Nissan and Jatko Ltd. top management

Fig. 6.22 Dr. Lev Klyatis (right) with Grom-Maznichevsky LI academician of Ukrainian Academy of Science by Cobo Center (Detroit), where SAE World Congress celebrated 100 years SAE (the author was one of 100 winners of natural leather jacket among near 50,000 attendees of this congress)

One more example involved a joint symposium held in 2012 by the partnership of the US Department of Defense and the US Department of Transportation and Industry (see part of program in Fig. 6.24) entitled " Road Map to Readiness at Best Cost for Improving the Reliability and Safety of Ground Vehicles." The author was invited as a tutor and speaker for (two presentations). Among the participating organizations were the US Army; the National Highway Transportation Safety Administration (NHTSA) (Department of Transportation); Volpe Center, Vehicle Research; Honda Motor Company; Active Safety Engineering, Raytheon Integrated Defense Systems; University of Alabama; ReliaSoft; Alion Science and Technology Corporation; and the Office of Secretary of Defense. This workshop provided an excellent opportunity for discussing the implementation of the author's ideas, approaches, and technologies for reliability/durability testing and efficiency prediction.

SAE International's Annual World Congress & Exhibition, claimed to be the largest technology meeting in the world, is directed toward all areas of mobility engineering and technology.

Attending are from 11,000 to 50,000 professionals from many countries each year. Their backgrounds vary through such disciplines as automotive, aerospace, off-highway, farm machinery, electrical, and electronic engineering. In addition to engineers, all levels of management from these industries can be found in attendance. Each year, presented papers are published by SAE International. The Event Guide for the Congress usually consists of 220–250 pages, including general information, special events and networking opportunities, ride and drive, management programs, chat with the experts, over 100 technical sessions, committee task force

RMS in a Systems Engineering Environment
A Government-Industry Conference

Program   DAU-West, San Diego, CA   October 12, 2006

7:30 – 8:30	Registration & Breakfast	All Attendants
8:30 - 8:45	Second day Opening	Dr. Russell Vacante, President, RMS Partnership
8:45 – 9:15	Fourth Keynote	CDR Al Desmarais, Southwest Regional Maintenance Center, Business Office
9:15 – 10:00	Third Panel	Enablers for RMS: Leader Martin Sherman, DAU West Region, Logistics Department Chair Tyrone Jackson, SRS Technologies, Performance-Based RMS Standards – a New Approach Eric Hausner, Northrop Grumman Unmanned Systems, Supportability Requirements for Net-Centric Systems Jim Rodenkirch, Systems Engineering LLC, Reliability Considerations in the Architecture of a System of Systems Brett Andrews, DAU Midwest Region, Sense and Respond and its Application to RMS(L)
10:00 – 10:15	**BREAK**	
10:15 – 11:45	Third Workshop	Break Out Rooms
11:45 – 12:45	**LUNCH**	Registrants will go to nearby facilities.
12:45 – 13:15	Fifth Keynote	CAPT Fred Cleveland, Commanding Officer, Fleet Readiness Center South West
13:15 – 14:00	Fourth Panel	Support to the Forces: Leader: Tim Greene, Space Warfare Systems Command CAPT Mike Kelly, AIRSpeed James Schrope, NADEP North Island, Integrated In-Service Reliability Program(IISRP) Team Jim Kral, SSC Charleston, Distance Support Lev Klyatis, Eccol, Inc. Elimination of the Basic Reasons for Inaccurate RMS Predictions
14:15 – 14:30	**BREAK**	
14:30 – 16:00	Fourth Workshop Break Out Rooms	Panel Leaders
16:00 – 16:30	Conference Close-Out	Dr. Russell Vacante

Please see the RMS Partnership web site (www.rmspartnership.org) for further details.
Continuous Learning Point Certificates will be issued.

Fig. 6.23 A part of program, a government-industry conference, DAU West, San Diego, CA

and board meetings, awards and recognitions, seminars, and exhibits. The 2013 SAE World Congress Event Guide cited the author's achievements in the development of successful prediction of product efficiency, during a chat with the experts session. Since 2012 he has been a chairman and co-organizer of the technical session "IDM300 (IDM100) Trends in Development Accelerated Reliability and Durability Testing" [9] which was organized in direct result of his work in this area. One example can be seen from Figs. 6.25, 6.26 and 6.27.

The above is illustration of the author's solutions.

RMS PARTNERSHIP
Reliability, Maintainability and Supportability

A Road Map to Readiness at Best Cost for Improving the Reliability and Safety of Ground Vehicles

DAY 1 Program September 19, 2012

8:15 – 9:00	Registration & Refreshments	Networking Opportunity
9:00 – 9:10	Conference Introduction	Dr. Russell Vacante, President, RMS Partnership
9:10 – 9:40	Opening Keynote Address (DoT)	HON. David L. Strickland, Administrator, National Highway Traffic Safety Administration (NHTSA)
9:40 – 10:10	Sustaining Readiness at Best Cost	Walter B. Massenburg, Senior Director, Mission Assurance Business Execution, Raytheon Integrated
10:10 – 10:20	BREAK	Break Area
10:20 – 11:20	Opening Keynote Address (DoD)	HON Katrina McFarland, Assistant Secretary of Defense (Acquisition), Office of the Secretary of Defense (Acquisition, Technology & Logistics)
11:20 – 12:00	Terrain Roughness and Suspension Capability - The Keys to Improved Reliability	Brett Horachek, Program Manager at Nevada Automotive Test Center
12:00 – 12:45	LUNCH	Break Area
12:45 – 2:00	Panel #1 - A Futuristic View of Ground Transportation Systems	- Melvin "Mel" Downes, Chief, Weapons, Fire Control & SW Quality, Reliability & Safety Engineering, the Quality Engineering and System Assurance Directorate, Armament Research, Development and Engineering Center, Picatinny Arsenal, NJ. - Gary Ritter, Director, Center for Advanced Transportation Technologies, RVT-90 Volpe Center - Robert E. Wild, Tactical Vehicles & Tires, Research, Development and Engineering Center, Army Tank Automotive (TARDEC), USA
2:00 – 2:30	Early Testing and Evaluation Impact on Reliability and Cost	John Paulson, Senior Director, Engineering Development and Technology, General Dynamics Land Systems
2:30 – 3:25	Panel #2 - Technology Innovation (The Impact of New Technology on Ground Vehicle Reliability, Safety & Cost)	- Joseph Dunlop, Vice President, Test Operations, Calspan Corporation - Tracy V. Sheppard, Director, Automotive Directorate, US Army Test Center, Aberdeen Providing Ground, Md. - Greg Kilchenstein, Deputy Assistant Secretary of Defense, Maintenance
3:25 – 3:35	BREAK	Break Area
3:35 – 4:05	Ground system Reliability	Dr. David Gorsich, Chief Scientist, U.S. Army Tank Automotive Research, Development and Engineering Center (TARDEC), TACOM LLC
4:05 – 5:30	Panel #3 - Reliability & Safety Standards for Ground Vehicles	- James French, Committe Manager, Mission Assurance Standards Working Group - Dr. Lev Klyatis, Senior Advisor, SoHaR, Inc. - Christopher J. Bonanti, Associate Administrator for Rulemaking, National Highway Traffic Safety Administration - Chris Peterson, Reliability Consultant, H & H Environmental Systems, Inc.
5:30 – 6:15	Tutorial on Reliability Testing and Accurate Prediction of Reliability & Safety	Instructor, Dr. Lev Klyatis, Conference Room, various conference all locations
6:15 – 7:30	Social Mixer	RMS Partnership & Sponsor Hosting

Fig. 6.24 The first day's program of DoD, DoT, and industry workshop/symposium

Typically, the events like World Congresses, Annual Reliability and Maintainability Symposium (RAMS), Quality Congresses, and similar gatherings include tours to nearby major industrial companies.

One surprising observation by author from these tours was how even large high technology companies, such as Thermo King, Lockheed Martin, Boeing, Detroit Diesel, and others, were still conducting (and continue using) separate vibration

Fig. 6.25   One example of author's presentation at the chat with experts of SAE World Congress

testing, temperature/humidity testing (and called them environmental testing), and corrosion testing with only chemical pollution of components in chambers with simulation of only a laboratory degradation of element at a time.

These companies still do not understand what is necessary for successful prediction of product efficiency. When Jay Seshadri, Vice President, International Engineering for Thermo King Corporation invited author to help them improve their product reliability, he agreed to consult with them to improve their expertise and their equipment for durability and reliability testing. Some of this work is described in greater detail in his published books [1, 2, 5].

Fig. 6.26 Lev Klyatis, chairman of technical session IDM300 Trends in Development Accelerated Reliability and Durability Testing, SAE 2013 World Congress, introducing a speaker—Zigmund Bluvband, President and CEO ALD Group (Israel)

The author observed similar situations during visits to the NASA Langley Research Center with others experts from the SAE G-11 Division in aerospace standardization to the Lockheed Martin plant during the ASQ Congress in Denver and many other industrial companies (see Chap. 2 of this book). In most all of these instances, he was surprised to observe that there was little implementation into the process reliability/durability testing and successful prediction of their product's efficiency. Even while he has not been able to visit these companies recently from their representative presentations, it is evident that their implementation of advanced technologies of efficiency prediction and accelerated testing is moving forward very slowly.

For example, the magazine *Aerospace Testing International*, which contains mostly information about flight testing, only articles about wind tunnel testing, vibration testing of some components, or about sensors are to be found.

If the reader will open the magazine *Aerospace & Defense Technology, The Engineer's Guide to Design & Manufacturing Advances* of SAE International for 2021–2022, he/she will not find any article or information about development technology of efficiency prediction or technology of reliability or durability testing or any type of other serious testing of the product. This demonstrates the increasing gap between design speed of development and testing development as was demonstrated in Chap. 5 (See Fig. 5.4). This also demonstrates the level of improving engineering product efficiency. For example, magazine issue for April 2022 consists of the articles in the following contents:

Fig. 6.27 Dr. Lev Klyatis during SAE Fellow reception and dinner, Detroit, Marriott Hotel

- Artificial intelligence
- Aerospace materials
- Aircraft electrification
- Mission critical systems
- Cabining and connectivity
- RF and microwave technology
- Show reviews
- Tech briefs

But qualified professionals understand that advanced engineering development needs corresponding testing and prediction development.

Over a period of 5–7 years after coming to the USA, the author has attended and provided numerous presentations, at the Annual Reliability and Maintainability Symposium (see Fig. 4.32 in [1]) and at several IEEE Workshops on Accelerated Stress Testing (see Fig. 6.28).

SPECIFICS OF ACCELERATED RELIABILITY TESTING

♦ LEV KLYATIS

♦ Habilitated Dr.-Ing., Sc.D., PhD

♦ Professor Emeritus

Lev Klyatios   ASTR 2009 Oct 7 – Oct 9, Jersey City, NJ   Abbreviated Title Page 1
July 20, 2017

Fig. 6.28 The title page of visuals during the presentation at the IEEE ASTR Workshop (2009)

These events included discussions with colleagues from different countries regarding the problems in testing and latest ideas for successful efficiency prediction. Through them all, he has observed how many attendees did not understand the importance and usefulness of the concepts and strategy of accelerated reliability/durability testing for successful efficiency prediction. Rather their presentations related only to components of successful prediction methodologies, leaving the complete development of methodology simulation, reliability testing, and prediction for the service life to others and other times.

But it also must be reiterated that successful implementation of new ideas and technologies depends not only on the strength of the ideas or technology but on the benefits they will provide in terms of testing effectiveness and the cost benefits that will be accrued over the long financial term. All this being said, it is somewhat encouraging to see how the advanced research centers over the world are moving in the direction of new test chambers development that includes multiple variables. Canada developed one such test chamber, described in further detail in the author's book *Successful Prediction of Product Performance* [4]. The level of testing development does not change seriously until now as we can see from the programs of the above events in the years 2021 to 2022.

One of the author's popular presentation was "Establishment of Accelerated Corrosion Testing Conditions" that was given at a RAMS (Annual Reliability and Maintainability Symposium: The International Symposium of Product Quality and Integrity), in January 2002. Following the presentation many professionals from industrial companies waited to speak with him and to discuss how they could implement his original approach to corrosion testing in their organizations. He recommended that they implement an advanced corrosion test methodology for their entire assortment of mobile products. Such an approach could account for all of the elements that could damage components and protective films. This was particularly important as many of these products were combined with mechanical wear or exposure to solar radiation. This is why the paper illustrates the modern essential of accelerated corrosion testing as a component of ART/ADT (see Fig. 5.6).

Early in his career in America, the author's presentations and published papers were aimed at different areas of industry. However, since the author had only recently arrived from Russia, his oral English abilities were limited. Consequently, his presentations at symposiums were often just written.

However, he was invited to give a presentation at an IEEE Workshop that included a visit to NASA's Jet Propulsion Laboratory (JPL). This resulted in a particularly memorable experience resulting from an IEEE Workshop that included a visit at their facilities. For more information on this visit, see its description in Sect. 2.3). Eventually, this first exposure in the American engineering community led to further invitations for other presentations.

During this time, his presentations were included in the conferences as basic information. For example, below one can see title page of his presentation *Specifics of Accelerated Reliability/Durability Testing* that was given at an IEEE Workshop Renewable Reliability, ASTR 2009, Jersey City, NJ Hyatt Regency, USA, 09/16/2009 (Fig. 6.28).

Such opportunities were important because they gave the author the opportunity to advance his ideas about the implementation of successful prediction components. This exposure led to his providing services as a consultant and seminar instructor for companies and organizations (especially large ones). For example, the following seminar for Ford Motors Company described an innovative approach to accelerated reliability testing (seminar contents shown below).

Trends in Development of Accelerated Reliability Testing
(Content of seminar for Ford Motor Company, April 2011)
   Instructor Dr. Lev Klyatis (Director of Quality and Reliability ERS Corporation, Head of Reliability Department ECCOL, Inc.)

- *Development techniques and equipment for more accurate simulation of real-life input influences*
- *Step-by-step development of less expensive equipment for simultaneous combination of real- life basic input influences*
- *Development accelerated reliability testing, which offers the possibility to obtain directly the information for accurate prediction of reliability*
- *Rapid obtaining of accurate information for analysis the reasons of degradation mechanism and failures*

- *Development the product quality through accelerated reliability testing. Its specific*
- *Development of accelerated analysis of the climate influence on the new product reliability*

What is ART development? It is when you have more opportunities for:

- *Rapid finding of product elements that limit the product's quality and reliability*
- *Rapid finding of the reasons for the limitations*
- *Rapid elimination of these reasons*
- *Rapid elimination of product overdesign (cost saving) to improve the product quality and reliability*
- *Increasing product quality and reliability, therefore resulting in a longer warranty period*

It is essential to simulate real-life input influences on the product accurately for ART. If we cannot simulate real-life influences accurately, we cannot perform ART and rapidly improve the product reliability.

The author is working in this direction and describes below how one can use it. One can use the above for development technology of ART:

- *Determination the failures that limited the product reliability and quality*
- *Finding the location and dynamic of mechanism development of the above failures (degradation)*
- *Finding the reasons of the above failures*
- *Elimination of these reasons*
- *Increasing of product's reliability and quality*
- *Increasing the warranty period of the product.*

For the above ART implementation, one needs accurate simulation real life influences on the actual product.

A very important aspect of expanding the knowledge and implementation of successful efficiency prediction has been internationally involved by Russia, the USA, and world organizations in efficiency and testing. Participation in this global experience helped the author to better understand the difference between Russian, American, and other nation philosophies and practice relating to engineering, efficiency, and its components. Unfortunately, it also revealed that many scientists, including many who had emigrated from the former Soviet Union, still would not adopt the differing cultures in reliability and durability testing and prediction even though they worked for many years in America.

Still, this author firmly believes that working with, and having discussions with, other world experts expands one's knowledge and opens the possibilities for improving and implementing modern technologies to create accurate simulation of real-world conditions, reliability testing, and successful prediction.

To date the most important method of disseminating these theories has been through seminars with primarily large industrial companies. Through them and visiting many companies and research centers, it has been possible for the author to study a variate of reliability/durability testing programs and develop ways for his approach to be more easily implemented in practice. Examples include seminars for Thermo King Corporation in 2010 that brought about significant changes in their operations. These experiences were described in [3].

The experience gained from these activities proved to be instrumental in helping the author to understand the direction that needed to be developed to advance the implementation of successful prediction of product efficiency and the need take into account the specific nature of and cultural differences of different companies and organizations in Western, Eastern, and Asian countries.

Other forums that added to the author's knowledge included numerous updated presentations from 2010 through 2017, working as an ASQ Council of Reviewers member, reviewing numerous book proposals and books for different world publishers, including drafts of books, book reviews, papers review for SAE World Congresses and ASQ Congresses, and work for the American federal government. All of these activities were involved in advancing the development and implementation of the successful prediction of product efficiency.

The questions posed during these presentations also served to provide valuable feedback enabling better understanding of how to further the development of new directions for the successful prediction.

The importance of the author's ideas in the development of successful prediction was further spread through his presentations that were organized by the SAE Metropolitan Section (see example on Fig. 6.32) and NAFA Fleet Management Association (example on Fig. 6.29).

The author does not know directly how wide in the world results of his new scientific solutions were implemented. However, there is information that over 3000 large libraries (public libraries, universities, industrial, and other companies) use his publications. One can see the example below that came from the ExxonMobil Corporation library, which had 59 of the author's publications in 2021.

Lev Klyatis Everything ▼ ADVANCED SEARCH

All items ▼ that contain my query words ▼ **as author/creator** ▼

0 selected **PAGE 1** 59 *Results*

PAGE 1

₁**BOOK**

Accelerated reliability and durability testing technology

©2012

Klyatis, Lev M.

View Online Availability

₂**BOOK**

Trends in development accelerated testing for automotive and aerospace engineering

2020

Klyatis, Lev M.

View Online Availability

₃**BOOK**

Accelerated quality and reliability solutions

2006

Klyatis, Lev M. Klyatis, Eugene L.

;

View Online Availability

METROPOLITAN SECTION

*** DOUBLE MEETING ***

*Methods to Maximize Quality, Reliability, Durability, Maintainability, and Cost-Effectiveness
In Product Development*

Lev Klyatis
Professor Emeritus, Dr-Ing, ScD, PhD, SAE Fellow, ASQ Fellow

This presentation will provide a description of the scientific and technical bases to be considered in product development, including: definitions of reliability, durability, maintainability, and cost-effectiveness. Lev will offer an analysis of current quality, reliability, and maintainability practices to achieve these ends and will discuss the basic reasons for the negative aspects of the current situation; what steps can be taken to eliminate these negative aspects; and the role of prediction and accelerated reliability testing to maximize quality reliability, maintainability, durability, and cost-effectiveness.

Attendees will receive a copy of Dr. Klyatis's recent book
Successful Accelerated Testing
⸬⸬⸬
*The Automotive Equipment Acquisition Process
Things You Need To Remember*

Edward L. Anderson, PE

This presentation on the acquisition of fleet automotive equipment will focus on helping smaller agency and authority fleets to recognize the need to specify and define their requirements with respect to many of the details related to equipment acquisition that are beyond vehicle specifications. Items such as training, manuals, delivery requirements, spare parts and PM kits, warranties, and other contractual provisions also need to be considered. A checklist will be provided to aide in writing and reviewing acquisition specifications, along with a list of resources available for those writing specifications.

Hope you will be able to join us!

Thursday, November 05, 2009
Registration: 11:30am – 12:00pm

Lunch: 12:00pm – 12:45pm
Presentation 12:45pm – 2:00pm

Meal choice: Roast Beef, Chicken, or Salmon
$25 (Members) / $30 (Guests) / $10 (Students)

Holiday Inn Carteret
1000 Roosevelt Avenue, Carteret NJ 07008 (732-541-9500)

DIRECTIONS: Take the NJ Turnpike North or South, to Exit 12 Carteret, Rahway (milepost 95.9). After toll booth, bear right at traffic light at the end of exit ramp. Holiday Inn is just past the light on the right.

Registration required by October 29, 2009
Contact: Lois Roth 201-216-2363/E-mail: lroth@panynj.gov
Payment: cash or check at the door
Attendance limited to 50

SAE Metropolitan Section

Fig. 6.29 Example organized by SAE Metropolitan Section from the lecture of Dr. Klyatis

₄**BOOK**

Reliability prediction and testing textbook

2018

Klyatis, Lev M., author. Anderson, Edward L., 1945- author.

;

View Online Availability

₅**BOOK**

Trends in Development of Accelerated Testing for Automotive and Aerospace Engineering

2020

Klyatis, Lev M

View Online Availability

₆**BOOK**

Reliability Prediction and Testing Textbook

2018

Klyatis, Lev M ; Anderson, Edward L

*This textbook reviews the methodologies of reliability prediction as currently used
in industries such as electronics, automotive, aircroft, aerospace,...*

View Online Availability

₇**BOOK**

Accelerated Reliability and Durability Testing Technology

2012

Klyatis, Lev M ; Klyatis

*Learn how ART and ADT can reduce cost, time, product recalls, and customer
complaints This book provides engineers with the techniques and tools they need to...*

View Online Availability

BOOK

8

Accelerated Quality and Reliability Solutions

2005

Klyatis, Lev M ; Klyatis, Eugene L.

Drawing of real-world issues and with supporting data from industry, this book
overviews the technique and equipment available to engineers and scientists to...

PAGE

View Online Availability

9 ARTICLE

A Methodology for Selecting Representative Input Regions for Accelerated Testing

Klyatis, Lev ; Walls, Lesley

Quality engineering, 2004-01-04, Vol.16 (3), p.369-375

This article describes an approach to the methodology of selecting a representative
region from the multitude of input influences based upon input (output)...

PEER REVIEWED

View Online Availability

10 REPORT

Implementation of Positive Trends in Reliability and Durability Testing and Prediction

2021

Klyatis, Lev

This paper considers the situation with implementation of positive trends in reliability
and durability testing and prediction. Professionals often obtain...

View Online Availability

REPORT
11

Analysis of Basic Directions of Accelerated Testing Development

2020

Klyatis, Lev

This paper provides the author's analysis of accelerated testing development. It considers four basic directions related to this development, specifically: 1)...

View Online Availability

REPORT
12

Basic Positive and Negative Trends in Development Accelerated Testing

2019

PAGE

Klyatis, Lev

This paper will discuss the problem of current trends in development accelerated testing. It will consider advantages (positive aspects) and disadvantage...

View Online Availability

....
....

58. BOOK CHAPTER

Chapter 3 - Accurate prediction of reliability, durability, and maintainability on the basis of useful accelerated reliability testing results

Lev M. Klyatis ; Eugene L. Klyatis

Accelerated Quality and Reliability Solutions, 2006, p.285-361

This chapter deals with the accurate prediction of reliability, durability, and maintainability by using the results of useful accelerated reliability testing...

View Online Availability

59. ARTICLE

A Methodology for Selecting Representative Input Regions for Accelerated Testing.

Klyatis, Lev; Walls, Lesley

Quality Engineering, 2004, Vol.16 (3), p.369-376

View Online Availability

6.6   Implementation Through Published Citations

Following are some copies of publications throughout the world with the author's publications, citations, and references (the author obtained this information several years ago through the Internet) that may be of assistance in successful implementation of reliability/ durability testing and successful prediction.

These citations were used in the following areas:

- Improving various product reliability issues in transportation refrigeration
- Marine energy converters
- Model aircrafts and launcher controllers
- Statistical processing of accelerated life cycle data
- Development of product models
- Durability testing of various products
- Improving quality test and reducing customer complaints
- Improving wear and life characteristics of composite products
- Development of various aspects of reliability
- Development of various aspects of reliability testing
- Development of reliability assessment
- Development of electronic systems reliability
- Renewable and sustainable energy

Some citations follow here:

(a) Joseph Mannion. Project: Computer Integrated Testing for the Transport Refrigeration Systems Industry. Mayo Institute of Technology. DATE: 7/9/2000:

"One important aspect of Accelerated Testing is the step-by-step strategy that can help obtain accurate and rapid initial information as results of product testing for reliability problem solving [6-7] (Lev M. Klyatis, 1999)."

(b) Philipp R. Thies, Lars Johanning, George H. Smith. Toward component reliability testing for marine energy converters. EMPS—Engineering, Mathematics, and Physical Science Renewable Energy Research Group, University of Exeter Cornwall Campus, Treliever Rd, Penryn, TR10 9EZ, UK. Preprint submitted to Ocean Engineering May 26, 2010:

"The type of test can be further distinguished, depending on how accurate the field loads are replicated and to what extent they are accelerated (Klyatis and Klyatis, 2006):

- *Field testing of the actual system under accelerated operating conditions*
- *Laboratory testing of actual system through physical simulation of field loads*
- *Virtual (computer-aided) simulation of system and field loads."*

(c) Allen Zielnik. Atlas Materials Testing Technology LLC. Validating Photovoltaic Module Durability Tests. Solar America Board for Codes and Standards. Ametek Corporation. www.solarabcs.org. July 2013:

"Klyatis (Klyatis, 2012) further points out that the vast majority of literature references to reliability tests are actually true durability tests, or stop short of failure at a finite duration as a qualification test. Despite extensive use of the term "reliability," most of what is being done in the PV industry at best falls under the realm of trying to assess "durability."

"As Klyatis notes, many types of environmental influences act on a product in real life. Some influences are studied, but many are not. The factors responsible for the influences of environmental stresses in the field are very complicated.

One of the most complicated problems is the integrated cause-and-effect relationship of different factor steps including stress influences, effect on output parameters, and degradation. In accelerated testing, we have two primary means of test acceleration. The first is overstress where one or more levels of a stress condition (such as temperature cycling) are applied at levels in excess of the intended normal service use. The test results are then used to extrapolate estimated performance at the normal stress level. Care must be taken not to exceed the stress strength of the product and induce unrealistic failure."

(d) Lisa Assbring, Elma Halilović. Improving **Ikea's** Quality Tests and Management of Customer Complaints—Kitchen Fronts and Worktops. Master of Science Thesis MMK. 2012:25 MCE 275 KTH Industrial Engineering and Management Machine Design SE-100 44 STOCKHOLM. Sweden:

"With AT, results that would require years of field testing may be obtained in days or weeks (Klyatis & Klyatis, 2005).
There are three general methods of accelerated testing. The first method is to test a product under normal field conditions and subject it to the same impact that would happen in real life (Klyatis & Klyatis, 2005). When the hinge on a kitchen front is tested, it is subjected to the same impact of being opened and closed that would happen during normal use in a kitchen.

The second method is to use special equipment in a laboratory to simulate the influences on a product (Klyatis & Klyatis, 2005). An example of this is when the wear and tear of years of usage of a worktop is simulated using a tool that causes abrasion on the surface. Since the influence on the product is simulated with a tool, the conditions do not correspond to normal field conditions. Accurate test results require accurate simulation of the field input.
The third method is to use computer software to simulate the effect on the product. The accuracy of this method depends on how accurately both product and the influences on the product can be simulated (Klyatis & Klyatis, 2005).

The approach of AT is to either accelerate the use-rate or the product stress. Use-rate acceleration can be used when a product is normally not continuously in use, and it is assumed that its lifetime can be modeled in cycles. These cycles are accelerated during testing and the product's lifetime estimated. This is sometimes referred to as ALT (Klyatis & Klyatis, 2005). For example, the expected number of times the kitchen front will be opened during its lifetime can be calculated, and by accelerating these cycles the effect they will have on the hinge can be tested."

"In accelerated stress testing the stress factors that cause product degeneration are accelerated. For example, temperature or concentration of chemicals may be increased in comparison to normal usage of the product. In the wear test of the worktop both the stress caused by the abrasion and the cycles are accelerated. Stress acceleration is often used to identify stress limits and design weaknesses. Combined stresses, such as temperature and vibration, are often especially efficient in accelerating failure of a product (Klyatis & Klyatis, 2005)."

"The equipment used during testing will also affect the results, as it can often only simulate one type of input while the product during normal conditions may be exposed to several simultaneously (Klyatis & Klyatis, 2005)."

(e) Gino Rinaldi. NSERC Research Fellow. A Literature Review of Corrosion Sensing Methods. Technical memorandum. Defense Research and Development, Canada.   September   2009.   *cradpdf.drdc-rddc.gc.ca/PDFS/unc102/p533953_A1b.pdf*

"A. 13.40 Klyatis (2002): Establishment of Accelerated Corrosion Testing Conditions

The author of this work developed techniques for accelerated corrosion testing (ACT) and implemented them. The main goal of ACT is rapid product improvement, lower warranty costs, lower life-cycle cost, and improved reliability through accelerated corrosion testing. [168]"

(f) Accelerated stability testing: Topics by Science.gov. Sample records for **accelerated stability testing.** Conditions of environmental accelerated testing (Citation 2). Microsoft Academic Search:

"Strategic and tactical bases are developed for conditions for environmental accelerated testing of a product using physical simulation of life processes. These conditions permit rapid attainment of accurate information for reliability evaluation and prediction, technological development, cost effectiveness and competitive marketing of the product, etc.(L. M. Klyatis)."

(g) Youji Hiraoka and Katsumari Yamamoto, Jatko Ltd. (Japan); Tamotsu Murakami, University of Tokyo; Yoshhiyuki Furukawa and Hiroyoki Sawada, National Institute of Advanced Industrial Science and Tech. (Japan). Method of Computer-Aided Fault-Tree Analysis (FTA) Reliability Design and Development: Analyzing FTA Using the Support System in Actual Design Process. SAE 2014 World Congress. SAE Technical paper 2014-01-0747:

"In reliability engineering the ART/ADT strategy components is advocated by Klyatis Lev [21]. This is a method of product quality guaranteed by testing process that uses FMEA and FTA to investigate the testing conditions of ART/ADT."

6.7   Implementation Through Published Reviews

There are many published reviews to published Dr. Klyatis's 12 books (6 books in English), over 250 articles, and papers.

The author is familiar with over 20 reviews in English. Let us show four of these reviews published in journals (partly) (Figs. 6.30 and 6.31).

"Book Report
ACCELERATED QUALITY and RELIABILITY SOLUTIONS

Lev Klyatis & Eugene Klyatis (Eds.)
Edited by ELSEVIER, EUR 136

Drawing of real-world issues and with supporting data from industry, this book overviews the technique and equipment available to engineers and scientists to identify the solutions of the physical essence of engineering problems in simulation, accelerated testing, prediction, quality improvement and risk during the design, manufacturing, and maintenance stages. For this goal the book integrates Quality Improvement and Accelerated Reliability/Durability/ Maintainability/Test Engineering concepts.

The book includes new unpublished aspects in quality: – complex analysis of factors that influence product quality, and other quality development and improvement problems during design and manufacturing; in simulation: - the strategy for development of accurate physical simulation of field input influences on the actual product – a system of control for physical simulation of the random input influences – a methodology for selecting a representative input region for accurate simulation of the field conditions; in testing: – useful accelerated reliability testing (UART) – accelerated multiple environmental testing technology – trends in development of UART technology; in studying climate and reliability; in prediction – accurate prediction (AP) of reliability, durability, and maintainability – criteria of AP – development of techniques, etc."

4   Reliability Review. ASQ. THE R & M ENGINEERING JOURNAL

Volume 26, Number 4, December 2006, *Page 29*
RR BOOKSHELF

"ACCELERATED QUALITY AND RELIABILITY SOLUTIONS
by Lev M. Klyatis and Eugene L. Klyatis. Elsevier, Oxford, UK, 2006

Most professionals entering the workforce today are required to analyze situations, identify problems, and provide solutions for improved performance. This book is particularly salient for engineers and managers for practical improvements in simulation, accelerated reliability testing, and reliability/maintainability/durability prediction. The content of this book is structured with five chapters:

1. *Accurate physical simulation of field input influences on actual product*
2. *Useful accelerated reliability testing performance*
3. *Accurate prediction of reliability, durability, and maintainability from reliability testing results*
4. *Practical quality development and improvement in manufacturing and design*

2 Book Review: Reliability Prediction and Testing Textbook

Lev M. Klyatis, Edward L. Anderson

Reviewer: Russell A. Vacante, Ph.D.

The Journal of RMS in Systems Engineering Winter 2019–Spring 2020

"During the early pages of this textbook the two authors Lev Klyatis and Edward Anderson, state, "without accurate simulation of real-world conditions, test results may be very different from the real world." From my perspective this is the cornerstone theme for all reliability modeling and prediction methodologies discussed by the authors. Their focus on this theme causes me to reflect on my early career attempts to gather "real-world" failure data from the field. Capturing this data was only partially successful given the limited human resources available to do so and the spotty communication network existing at the time. Back to the book!

What the authors are saying is that "accurate physical simulation of "field data" is a necessary first step towards achieving reliability modeling and predictions. To back this observation up, the authors provide a brief historical and acute technical insight into the various past and present reliability models, mythologies and testing procedures. For example, they address the background and use of MIL-HDB 217, the Bellcore/Telcordia Predictive Method, Physics of Failure (for electronic components) approach, and Life Testing Methodology. The authors underscore how they have been used, their attributes, and most importantly why they, for the most part, have only met with limited success in their application for predicting product performance.

The graphic on page 30 of the textbook nicely demonstrates that "reliability" is only one of many factors that helps determine a product's technical performance in the field. Factors such as safety, durability, and life cycle cost, among others, play an important role. While reliability may be only one of the factors that help to determine a product's technical performance the above-mentioned other factors do impact the overall reliability of a product. The trade-offs that are made pertaining to safety, maintenance and cost when developing a life cycle strategy have been known to both enhance and deteriorate product reliability. From a system engineering perspective, they are all interrelated. The design and development teams may find it in their interest to reduce the robustness of a product's inherit reliability by having the end user organization maintain product availability through the use of spares.

The use of accelerated reliability/ durability testing, ART & ADT respectfully, comes close to reflecting a product operating environment when periodic field data and laboratory simulation is combined. In this manner the field input has direct input on the variables such as output, vibration, tension, loading etc. On page 83 of the textbook the authors provide an excellent example of numerous inputs that should be considered to perform reliability predictions. This chart is a handy reference designed to make us mindful of real-world influences that will help lead to more accurate reliability predictions.

In summation, this book deserves high marks in providing an overview of the reliability methodologies and tools used in the past and at present. This book is a must read for engineers and managers who are interested in improving their engineering culture.

The most important take-away for me is that this book, from cover to cover, holds true to its opening theme: "without accurate simulation of real-world conditions, test results may be very different from the real world." The absence of real-world data in the previous reliability modeling and prediction methods is a flaw from which they cannot escape. Recent computer and related technologies have made the task of acquiring "real-world" data much easier and timelier. No testing is complete without it. As the authors have succinctly stated, without it the number of vehicles recalled, due to poor reliability-performance will continue to grow. Engineers - take the observation of the authors seriously. Make certain that you incorporate real-world data when doing accelerated reliability/durability testing."

Fig. 6.30 The book review, published in the "Journal of RMS in Systems Engineering," Winter 2019–Spring 2020

Total Quality Management
Vol. 17, No. 7, 959–960, September 2006

R Routledge
  Taylor & Francis Group

Book Review

Accelerated Quality and Reliability Solutions
Lev M. Kylatis & Eugene L. Klyatis.
Oxford, Elsevier, 2006, ISBN 0-08-044924-7, 544 pp. US$150/€136.

A new and very useful book on accelerated reliability testing is now available from Elsevier. The book focuses on the accurate depiction of field influences on a product's performance and how to simulate these in a laboratory environment. Whereas most books on accelerated testing concentrate on the mathematical aspects of analyzing the data, this book details how to develop the test to accurately reflect field usage, then how to duplicate this usage in actual laboratory testing. This information is sorely needed in industry today because of the need to design and develop high reliability components and systems in an ever-shorter time span in order to get the product to the marketplace before the competition.

Many companies today still rely on the test-analyze-test method to 'grow' reliability. There is much literature still being written on the outdated concept of reliability growth. It is still not generally recognized that a robust design, i.e. a design whose performance is insensitive to the variation in environments and usage conditions of the customer, must be developed with the initial design. Thus, there is one and only one chance to test the design in a representative, accelerated fashion to prove high reliability. Dr and Mr Klyatis's book provides valuable information on the how-to side of analyzing the inputs to identify the right factors that influence reliability, and then developing the accelerated test that duplicates those factors influence in the laboratory test.

The book is divided into five chapters. Chapter 1 details the strategy for developing an accurate physical representation of the influence of various field inputs. It reviews how to obtain accurate information from field, then selecting a representative input region to assure accurate field conditions in the test. It discusses the influence of climate on reliability, looking specifically at the influences of solar radiation, temperature, fluctuations of daily and yearly are temperatures, humidity rain, wind speed, and other atmospheric phenomena. The authors then show how each of these factors affects reliability of the products. The final section of chapter 1 details how to simulate these particular influences on the product using wherever necessary artificial media for these natural phenomena.

Chapter 2 deals with developing the specific accelerated reliability test for the products. It contains a detailed eleven-step process to develop the test, starting with collection of the field information and ending with using the results of the test for rapid, cost-effective improvements. It then discusses acceleration methods for solar radiation, chemical destruction, weathering, corrosion and vibration.

Chapter 3 show how to make useful reliability, durability, and maintainability predictions from the results of these accelerated reliability tests. The authors first review the mathematical basis for being able to predict from the results of accelerated testing. They then discuss the development of techniques to predict reliability without finding the analytic or graphical form of the failure distribution. This is followed by predictions using mathematical models with dependence between reliability and factors from the field and manufacturing. The chapter concludes with discussions on simple and multi-variate Weibull analysis, durability predictions, and predictions of optimal maintenance intervals and spare parts usage.

Chapter 4 expands from accelerated reliability testing to the use of accelerated methods for quality improvement and improvements in manufacturing. It reviews basic quality concepts, then shows how to use these in accelerated manufacturing improvement and design of manufacturing

1478-3363 Print/1478-3371 Online/06/070959–2 © 2006 Taylor & Francis
DOI: 10.1080/14783360600958120

Fig. 6.31 Published review in journal Total Quality Management and Business Excellence, Taylor & Francis Group, Volume 17, Number 7, September 2006, UK

960   *Book Review*

equipment. It then shows how to find those manufacturing factors that influence product quality and how to use them to improve product design.

Chapter 5 deals with the basic concepts of safety and risk assessment. It covers estimating risk; evaluating risk; performing hazard analysis; and managing risk. The chapter concludes with an introduction to human factors.

This book is an excellent reference text on accelerated testing. The book is of great benefit of the design engineer level. It covers in detail the many aspects of designing and developing an accelerated reliability testing not heretofore found in other texts. The only thing would have made this book even more valuable would have been the addition of problems at the end of each major section so that it could be used as a college-level textbook for engineers. Still, the book is an excellent addition to the library of every design engineer seeking to improve his/her design expertise.

Richard J. Rudy
Senior Manager, Product & Process Integrity (retired)
DaimlerChrylser Corporation

Fig. 6.31   (continued)

5. *Assessment of safety risk*

This updated and expanded edition presents many different solutions in an easy-to-follow structure: a description of the tool, its purpose and typical applications, the correct procedure is illustrated, and a checklist to help one apply it properly. The examples, forms, and templates used are general enough to apply to any industry or market.

The layout of the book has been designed to help speed your learning. The book is suited for employees working in Quality and Reliability fields in any type of industry. It includes a variety of data collection and analysis tools, tools for planning, and process analysis.

Important benefits of this book are:

New concepts which can help to solve earlier inaccessible problems during design, manufacturing, and usage
Can help dramatically reduce recalls
Develops a new approach to improving the engineering culture for solving reliability and quality problems
Prescribes a practice for improving reliability/maintainability/durability
This more than 500-page book is an excellent addition to the library of every industrial, service, processing, high tech, consultant, training company, as well as every engineer and manager.

Reviewed by:
Dak K. Murthy, P.E., C.Q.E.
Manager, Quality Assurance, NJ Transit and Chair of ASQ NY/NJ Metropolitan Section

References

1. Lev Klyatis (2012) Accelerated Reliability and Durability Testing Technology. Wiley
2. Lev Klyatis, Eugene Klyatis (2006) Accelerated Quality and Reliability Solutions. Elsevier, UK
3. Lev M. Klyatis, Edward L. Anderson (2018) Reliability Prediction and Testing Textbook. Wiley
4. Lev Klyatis (2016) Successful Prediction of Product Performance (quality, reliability, durability, safety, maintainability, life cycle cost, profit, and other components). SAE International
5. Lev M. Klyatis (2020) Trends in Development of Accelerated Testing for Automotive and Aerospace Engineering. Elsevier (Academic Press)
6. Lev M. Klyatis, Eugene L. Klyatis (2002) Successful Accelerated Testing. Mir Collection
7. 加速可靠性和耐久性试验技术 = Accelerated reliability and durability testing technology / Jia su ke kao xing he nai jiu xing shi yan ji shu = Accelerated reliability and durability testing technology. Author: 凯耶斯 (Klyatis, Lev M.) (俄)Lev M. Klyatis著 ; 宋太亮, 方颖, 丁利平等译 宋太亮, 方颖, 丁利平, ; Lev M. Klyatis.; Tailiang Song; Ying Fang; Liping Ding, Publisher: 国防工业出版社, Beijing: Guo fang gong ye chu ban she.
8. Karpov L (1990) Interview with Klyatis L. M., ScD (November 1990) High Quality Rigging for Testing Service. Editorial article. Journal Tractors and Farm Machinery. USSR Department of Automotive and Agricultural Industries
9. SAE 2013 World Congress. Achieving Efficiency. Event Guide. SAE International.
10. David Nicholls (2012) An Objective Look at Predictions – Ask Questions, Challenge Answers. Annual Reliability and Maintainability Symposium (RAMS) Proceedings
11. Lev Klyatis (2013) Development Standardization "Glossary" and "Strategy" for Reliability Testing as a Component of Trends in Development of ART/ADT." SAE 2013 World Congress (Detroit), paper # 2013-01-0152
12. Lev Klyatis (2013) Development of Accelerated Reliability/Durability Testing Standardization as a Components of Trends in Development Accelerated Reliability Testing (ART/ADT). SAE 2013 World Congress (Detroit). Paper #2013-01-151
13. Lev   Klyatis   (1985)   Book: Accelerated   Testing   of   Farm   Machinery. AGROPROMISDAT. Moscow, USSR
14. Lev Klyatis (April 2017) Why Separate Simulation of Input Influences for Accelerated Reliability and Durability Testing is Not Effective? SAE 2017 World Congress. Paper # 2017-01-0276. Detroit
15. Lev Klyatis (Spring 2016) Successful Prediction of Product Quality, Durability, Maintainability, Supportability, Safety, Life Cycle Cost, Recalls and Other Performance Components. The Journal of Reliability, Maintainability, and Supportability in Systems Engineering. RMS Partnership. Pp. 14–26
16. Lev Klyatis (2009) Accelerated Reliability Testing as a Key Factor for Accelerated Development of Product/Process Reliability. IEEE Workshop Accelerated Stress Testing. Reliability (ASTR 2009). Proceedings on CD. October 7–9, Jersey City
17. Lev Klyatis (2006) Introduction to Integrated Quality and Reliability Solutions for Industrial Companies. ASQ World Conference on Quality and Improvement Proceedings. May 1–3, Milwaukee, WI
18. Lev Klyatis, Lesley Walls (2004) A Methodology for Selecting Representative Input Regions for Accelerated Testing. Quality Engineering. Volume 16, Number 3, 2004, pp. 369–375. ASQ & Marcel Dekker
19. Klyatis, L. M (2002) Establishment of Accelerated Corrosion Testing Conditions. Reliability and Maintainability Symposium (RAMS) Proceedings. Seattle, WA, January 28–31, 2002, pp. 636–641
20. Klyatis, L. M. and Klyatis E (2001) Vibration Test Trends and Shortcomings, Part 1. *Reliability Review*. The R & M Engineering Journal (ASQ). Part 1, 2001 September, Volume 21, Number 3

Index

© The Editor(s) (if applicable) and The Author(s), under exclusive license to
Springer Nature Switzerland AG 2023
L. M. Klyatis, *Prediction Technologies for Improving Engineering Product
Efficiency*, https://doi.org/10.1007/978-3-031-16655-6

Printed in the United States
by Baker & Taylor Publisher Services